Forecasting Without Power

Why the Future Can Be Visible Before It Is Changeable

Ralph Clayton

Deluxe edition.

Technical Companion

This book is the narrative companion to a formal research program developed by the author in February 2026. The Volumetric Time Model (VTM), Operational Accessibility Theory of Selection (OATS), Agency Horizons, Future Access Without Presence (FAWP), and related constructs are presented here in accessible, story-driven form.The underlying mathematics, information-theoretic definitions, closed-form derivations, Monte Carlo experiments (E1–E8), null controls, bootstrap confidence intervals, and simulation code appear in the following technical preprints (all by Ralph Clayton, freely available on Zenodo)

Preprints (newest to oldest)

- **Forecasting Without Power: Agency Horizons, Control Collapse, and the Leverage Gap**
  Zenodo preprint, uploaded February 19, 2026.
  DOI: https://doi.org/10.5281/zenodo.18700470
- **Future Access Without Presence (FAWP): Operational Prediction Without Leverage in the Volumetric Time Model**
  Zenodo preprint, uploaded February 17, 2026.
  DOI: https://doi.org/10.5281/zenodo.18673949
- **The Agency Horizon: An Information-Theoretic Bound on Action at a Distance**
  Zenodo preprint, uploaded February 16, 2026.
  DOI: https://doi.org/10.5281/zenodo.18663547
- **Future Access Without Presence in the Volumetric Time Model**
  Zenodo preprint, uploaded February 15, 2026.
  DOI: https://doi.org/10.5281/zenodo.18651290
- **Steering in the Volumetric Time Model**
  Zenodo preprint, uploaded February 14, 2026.

DOI: https://doi.org/10.5281/zenodo.18639907

- **The Volumetric Time Model: Constraint-Conditioned Histories and the Operational Origins of Perceived Agency**
  Zenodo preprint, uploaded February 13, 2026.
  DOI: https://doi.org/10.5281/zenodo.18634216

These preprints contain the exact definitions

Definitions of key quantities, closed form solutions for representative channel classes, and the full experimental pipeline, including estimator choices, autocorrelation preserving nulls, robustness checks, thermodynamic bounds, functional viability boundaries, and quantum consistency constraints. A unified collection page and a maintained code archive for the OATS suite will be provided as project resources. Updates will be noted on the author site or the book resources page. Search for Ralph Clayton at Zenodo, it will also surface the latest versions and reproducibility artifacts. The narrative chapters translate these operational and geometric ideas into everyday language while preserving the underlying formal commitments. Readers seeking proofs, derivations, or replicable experiments should consult the technical preprints directly.

March 2026

Ducunt volentem fata, nolentem trahunt.
Fate leads the willing and drags the unwilling.
Seneca, Letters to Lucilius, 107.11.

FAWP

What is FAWP?

Future Access Without Presence (FAWP), is the condition in which you can see where events are heading while lacking enough timely influence to change the outcome in a meaningful way. It names a feeling that many people already know, even if they have never had precise language for it. You feel a relationship ending before the breakup happens. You sense that a job is dying before the official notice arrives. You know something is wrong in your body before a diagnosis is confirmed. You can see fragility in a market before the crash, yet still be unable to trade it cleanly, profit from it, or stop the damage. In all of these cases, the future is not completely hidden. It is partly visible. The problem is that visibility is not the same as control.

That is the core of FAWP. Prediction and power are not the same thing.

A lot of people instinctively assume that if they can understand a situation, they should also be able to steer it. In simple environments, that can feel true. When feedback is fast, when the system is small, when your action reaches the outcome quickly and clearly, knowing and doing can stay close together. But in real life, especially in modern life, the two often separate. Systems are noisy. Feedback is delayed. Action is diluted across many people, institutions, and layers of friction. By the time a pattern becomes clear enough to recognize, the window for changing it may already be narrowing or partly gone. FAWP is the name for that separation.

It does not mean magical foresight. It does not mean certainty. It does not mean fatalism, laziness, weakness, or passivity. It does not mean nothing can ever be changed. It means something narrower and more exact. It means there are situations where the information channel stays alive longer than the action channel. The world is still telling you what is likely to happen, but your ability to alter that future has already weakened.

This is why FAWP hurts so much. Ignorance is frustrating, but clarity without reach is worse. When you are completely blind, the pain comes from surprise. When you are in FAWP, the pain comes from recognition without leverage. You live through the bad future twice, first as anticipation, then as arrival. You feel the shape of what is coming, yet cannot fully connect your hands to the machinery in time to stop it.

One of the central ideas behind FAWP is the Agency Horizon. This is the practical boundary beyond which your influence becomes too delayed, too weak, too noisy, or too poorly linked to outcomes to matter reliably. Beyond that horizon, action may still exist in some abstract sense, but not in a way that gives you usable control. Another key idea is the Leverage Gap. This is the zone where prediction remains above the level of mere chance, while steering falls below meaningful detectability. In plain terms, you can still read the future better than random guessing, but you can no longer push it around effectively.

That pattern appears everywhere. In health, relationships, careers, finance, technology, institutions, and social life. Modern life may produce more FAWP than earlier eras because it gives people more signals, more dashboards, more commentary, more metrics, and more visibility into unfolding processes, while not giving them proportional control over those processes. The map gets brighter, but the steering does not improve at the same rate.

So FAWP is not just a technical concept. It is a way of describing one of the defining emotional experiences of the modern world. It explains why so many people feel that they can see what is coming and still cannot stop it. It explains why hindsight often feels cruel. It explains why late clarity can feel like personal failure, even when the deeper issue was structural timing rather than character. In the end, FAWP is about the painful gap between legibility and leverage. It is about the fact that the future can remain visible after your power to change it has already started to fade.

FAWP 101

FAWP stands for Forecasting Without Power.

It describes a situation that a lot of people already know in ordinary life, even if they have never had a name for it.

It is the experience of seeing what is coming but not being able to stop it.

You notice the signs. You read the pattern. You understand where things are heading. You may even predict the outcome correctly. But by the time the picture becomes clear, your ability to change it is weak, delayed, or basically gone.

That is FAWP.

This idea matters because people often assume that prediction and control are the same thing. They are not. Knowing what is likely to happen is not the same as having the power to change what happens. In fact, some of the clearest forecasts happen in exactly those situations where meaningful intervention is already too late.

Think about a few ordinary examples.

A person can see a breakup coming weeks before it happens. The signs are everywhere. The tone changes. The routines change. The warmth is gone. The future becomes easier to predict. But the power to fix it may already be collapsing.

A trader can see volatility building in a market. The danger becomes more obvious. The pattern gets sharper. Yet execution delays, slippage, and feedback problems can make that knowledge hard to use in time.

A city can see flooding risk rise. A hospital can see a surge coming. A company can see a failure building in a process. In all of these cases, the forecast may improve while the usable window for action shrinks. That is the heart of FAWP.

FAWP says there is a real difference between three things:

Prediction: what you can infer from the information you have now.

Influence: what your actions can still change.

Presence: whether you actually have access to the right moment, place, and channel to make that influence matter.

Those three things often get blurred together in everyday thinking. FAWP separates them.

You can have prediction without influence.

You can have awareness without leverage.

You can have information without presence.

That is not mystical. It is not fortune telling. It is not a claim that someone is literally downloading the future. It is a much simpler and more practical idea.

The world has structure.

Some systems carry clues about where they are heading. Large processes have inertia. Delays matter. Bottlenecks matter. Noise matters. Once enough momentum builds, the future can become easier to read even as it becomes harder to touch.

That is why FAWP is best understood as a structural condition, not a spooky one.

A good way to picture it is this:

Imagine you are standing on the shore, watching a huge wave form far out in the water. At first, it is small and hard to read. As it grows, its direction becomes more obvious. You become more confident that it is coming toward you. But confidence does not mean control. If you are trapped on the rocks, the better you can see the wave, the less that knowledge helps you change it.

That is FAWP in one image.

In the research version of the idea, this is framed in terms of two different kinds of coupling.

The first is predictive coupling. That means the information you have right now still contains real signal about what will happen later.

The second is steering coupling. That means your action right now still has detectable power to shape what happens later.

FAWP appears when predictive coupling remains alive, but steering coupling has dropped below meaningful influence.

In plain English: you can still read the future better than you can move it.

That distinction matters a lot, because modern life is full of systems where monitoring is easier than intervention.

Dashboards can update instantly while real decisions move slowly. Alerts can fire in seconds while institutions respond in days. Models can get more accurate while the intervention channel is clogged by delay, review, physical inertia, or simple scale.

So FAWP is not just a philosophical concept. It is also a practical one.

It helps explain why people often feel trapped in highly measured environments. We have more prediction than ever before. More sensors. More analytics. More warnings. More models. But more prediction does not automatically create more control. Sometimes it only makes powerlessness more visible.

That is why the name matters.

"Forecasting Without Power" captures a familiar modern experience. You are not blind. You are not irrational. You are not failing to notice reality. You may be noticing it perfectly well. The real problem is that the action channel is weaker than the information channel.

That idea also changes how we think about blame.

When people fail to stop something bad, we often rush to moral explanations. We say they ignored the signs. They were lazy. They did not care. Sometimes that is true. But FAWP asks us to consider another possibility: maybe the forecast arrived inside a structure where leverage had already decayed. That does not excuse every failure. But it gives us better language.

It lets us ask better questions.

How much of the future was actually visible?

How much of it was actionable?

How much delay existed between seeing and doing?

How much noise, friction, and lag stood between the decision and the outcome?

Those are FAWP questions.

And they matter because a society can become very good at measurement while becoming very bad at timely action.

In that kind of world, people may feel surrounded by warnings and yet unable to change much. The forecast gets sharper. The intervention window gets smaller. Knowledge rises as leverage falls. That is the paradox FAWP is trying to name.

So, in the simplest possible terms:

FAWP is the condition where the future becomes easier to predict than to influence.

You can see the crash coming, but not stop it.

You can read the pattern, but not redirect it.

You have access to the signal, but not to the moment where control still counts.

That is why FAWP matters.

It gives a name to one of the most common and frustrating experiences in modern life:

understanding what is about to happen while being too late, too far, too constrained, or too weak to change it.

Disclaimer

This is a science book written for everyday folks.

It is meant for curious readers who want to understand a big idea in plain, readable language. It is not a technical textbook. It is not a research monograph. It is not a specialist manual written for professional academics speaking only to one another. If you picked up this book because you want a clear, serious explanation of an important pattern in life, then you are exactly the kind of reader this book was written for.

The purpose of this book is not to impress experts with dense language. Its purpose is to make a difficult idea understandable, memorable, and useful.

The core subject of this book is simple to state, even if it has deep consequences. Sometimes people can see what is coming before they still have enough power to stop it. Sometimes the direction of events becomes visible before those events remain easy to steer. That experience appears in health, relationships, money, work, technology, and many parts of ordinary life. This book tries to explain that experience in a way that everyday readers can grasp without needing advanced training in mathematics, statistics, economics, physics, or philosophy.

That means the language in this book is intentionally accessible. Technical ideas are translated into ordinary speech wherever possible. Formal concepts are introduced only to the extent that they help the reader understand the larger argument. Some details that would matter in a full academic treatment are simplified, compressed, or left in the background on purpose. This is not because those details are unimportant. It is because this book has a different job. Its job is to make the central idea legible to normal human beings living ordinary lives.

This book cares much more about clarity than about performance.

In some scientific and academic settings, there is a habit of turning every idea into a wall of terminology, formal caution, and technical display. That style has its place. It can be useful when researchers are arguing about exact definitions, methods, boundary cases, or formal proofs. But it is not always the best way to introduce an idea to a broader audience. A reader should not need a graduate degree just to understand a pattern they may already have felt in their own body, their own bank account, their own family, or their own working life.

So this book takes a different path.

It starts from lived experience. It uses examples from everyday life. It explains technical ideas in ordinary language. It favors intuition before formalism. It tries to give the reader a clean mental map first, so that the structure of the idea becomes clear before the language around it becomes heavy. The ambition here is not to water down the subject. The ambition is to translate it honestly.

That distinction matters.

Simplifying an idea is not the same thing as disrespecting it. Clear writing is not shallow writing. Accessible writing is not lesser writing. Some of the most important scientific ideas become powerful only when they can be understood outside narrow professional circles. If an idea matters for human life, then everyday folks should be able to think about it too.

This book should therefore be read as serious popular science.

It is serious because it is trying to describe a real structure in human experience and connect it to a disciplined way of thinking about information, timing, feedback, action, and limits. It is popular science because it is written for a broad audience, not only for specialists. It is designed to be readable by intelligent non experts. It is meant to be followed, not decoded.

That also means this book is not a substitute for professional advice in high stakes situations. Nothing here should be treated as medical advice, psychiatric advice, legal advice, financial advice, or individualized professional guidance. When this book discusses examples involving health, markets, technology, relationships, or major life decisions, those examples are there to illuminate ideas. They are not there to replace expert judgment, clinical care, or domain-specific training.

The same goes for formal scientific treatment. Readers who want exhaustive proofs, full methodological detail, or specialist argumentation should understand that this book is not trying to serve that function. It is a doorway, not an archive. It is an explanation, not a complete technical record. It is meant to help readers think more clearly, not to bury them under professional machinery.

That is an intentional choice.

This book assumes that everyday folks are capable of understanding large ideas when those ideas are explained with respect. It does not assume the reader is foolish. It does not talk down to the reader. It simply refuses to confuse needless complexity with intelligence. The hope is that a reader can come to this book with curiosity, attention, and ordinary life experience, and leave with a new framework for understanding why seeing the direction of events is not always the same thing as having the power to change them.

If you are an academic or a technical reader, you are welcome here too. But this book was not built around academic pacing, academic style, or academic priorities. It was built around human readability. It was built to carry an idea across the bridge from formal thought into everyday understanding. It was built to help ordinary readers say, clearly and without shame, I know this feeling, and now I finally have language for it.

That is the spirit in which these pages should be read.
This is a science book for everyday folks.
It is here to explain and clarify.
To give a difficult idea a form that real people can actually use.
Everyday people like you and me.

Prologue

The most dangerous moment in a financial crash is usually not the day the crash becomes obvious.

It is the long stretch before that day, when the machine is still running, the lights are still on, the headlines are still smiling, and the people inside the machine can already feel that something has gone wrong.

That is one of the hardest truths in markets, and one of the hardest truths in life. A system can already be broken before it is allowed to look broken. A structure can already be failing before the official language changes. The risk can already be real while the price is still rising, the bonuses are still flowing, the neighbors are still celebrating, and the experts on television are still using calm voices.

This book begins there.

It begins with the strange and painful experience of seeing enough to fear what is coming, while lacking enough power to stop what is coming.

That experience has a name in these pages. Future Access Without Presence. FAWP. It sounds technical, but the feeling is familiar. It is the feeling of recognizing direction before you have leverage. It is the feeling of reading the shape of the future while your hands no longer reach far enough into the mechanism to change that shape cleanly. It is not magic. It is not prophecy. It is not hindsight dressed up as wisdom. It is a structural gap between what you can detect and what you can still control. Financial markets reveal that gap in the clearest possible way.

Markets are giant public theaters of belief. They are places where fear, greed, credit, momentum, habit, and imitation turn into numbers on screens. Every second, millions of decisions collapse into a price. That gives markets an aura of precision. A price looks clean. A chart looks objective. A line going up looks like proof. A line going down looks like judgment. People stare at these moving numbers and feel that the machine must be telling the truth because the machine is measurable.

But a market is not a truth machine. It is a pressure machine. It is a sorting machine. It is a crowd machine. It can reflect reality, distort reality, postpone reality, and then suddenly reintroduce reality with brutal speed.

That is why markets are such a powerful place to start if we want to understand FAWP.

In a market, it is possible to detect fragility long before the collapse is official. It is possible to see a structure becoming unsound while the public still calls it strong. It is possible to notice that a boom is being held together by leverage, borrowed confidence, and self referential optimism rather than by durable value. It is possible to see that the floor is thin while the ballroom above it is still full.

And yet the fact that you can see it does not mean you can stop it.

That is the split this book is about.

Think about what a market bubble really is. At first it does not feel like madness. It feels like intelligence rewarded. It feels like progress. It feels like a new era. Something starts rising. Early gains attract attention. Attention attracts more buyers. More buyers push prices higher. Those higher prices then become a public advertisement for the story itself. Look, people say, the market is confirming the idea. The rise becomes its own evidence. The proof is no longer outside the system. The proof is the movement of the system itself.

That is when danger gets dressed up as success.

Once that happens, skepticism becomes socially expensive. If prices keep rising, caution starts to look stupid. Discipline starts to look timid. Doubt starts to look old fashioned. A person who says this cannot last sounds boring while everyone else sounds rich.

Bubbles do not survive by hiding their excitement. They survive by making excitement feel rational.

Then the deeper changes begin.

Borrowed money starts doing work that judgment used to do. Risk gets cut into pieces, renamed, repackaged, and sold as opportunity. Rising prices cover over bad assumptions. Fragile structures look healthy because they are still producing gains. Incentives become aligned with continuation. More and more people need the story to remain true because their balance sheets, reputations, careers, pensions, or lifestyles are now tied to the story remaining true.

This is a very important point.

A fragile structure becomes hardest to challenge at the very moment it becomes most fragile.

That sounds paradoxical, but it is one of the central mechanics of modern financial life. When a system is only mildly distorted, people can still talk honestly about it. When a system becomes deeply distorted, too many people are being paid by the distortion. By then careers depend on it. Institutions depend on it. Political stability may depend on it. Household budgets may depend on it. The public story gets more confident precisely because the private reality is getting weaker.

That is why a crash can feel visible long before it feels stoppable.

The Great Depression is one of the clearest examples.

The late 1920s in the United States were full of confidence. Stocks climbed. Participation widened. Speculation spread beyond specialists into ordinary households. Credit expanded. People borrowed to buy shares. Paper wealth multiplied. The mood of the country became entangled with the movement of the market. The rise in prices was not just treated as a financial fact. It was treated as civilizational proof. It seemed to confirm intelligence, industry, modernity, national strength, and the promise of endless expansion.

To live inside that period was to live inside a story of ascent.

And that is what makes the danger so hard to see from the inside. The same movement that produces fragility also produces social permission. A rising market does not just make people wealthier. It changes what sounds reasonable. It changes who gets listened to. It changes the emotional cost of sounding cautious. It changes the difference between wisdom and pessimism.

Yet beneath all that enthusiasm, the structure was not as solid as it looked.

Debt was amplifying exposure. Speculation was outrunning restraint. Confidence was doing the work of analysis. Valuations were no longer anchored to sober expectations. The entire arrangement was becoming more dependent on continued belief, continued expansion, and continued participation. That kind of system can look glorious right up until the point where it becomes unable to carry its own weight.

This is the part that matters for our purposes.

People often talk about 1929 as if the crash arrived out of nowhere and then taught the world a lesson. But the more important truth is that the conditions of the crash were accumulating before the headlines changed. The system was becoming more brittle before the break. The danger was building before the official recognition of danger. Some people could see enough to be worried. What they could not do was force the entire structure to behave as though worry itself had authority.

That is FAWP in its most public form.

The future had become legible before it had become governable.

The signals were there, but the control channels were weak. Seeing that leverage was dangerous did not let you make millions of other people unborrow money. Seeing that speculation had outrun reason did not let you reverse the crowd. Seeing that prices had detached from reality did not give you a lever large enough to lower prices gently and safely. Even if you were right, the machine was bigger than your rightness.

And this is where the shame enters, in markets and everywhere else.

After a collapse, society rewrites the story. Suddenly everyone speaks as if the disaster had been obvious. People say the signs were there. They say somebody should have acted. They say people should have known better. They turn recognition into moral clarity after the fact.

But knowing and stopping are not the same act.

A person can know that a market is unhealthy and still be unable to stop the market. A family can know that debt is becoming dangerous and still be unable to exit without immediate pain. An analyst can understand the weakness in an institution and still be too low in the hierarchy to redirect the institution. A central bank can see overheating and still face political and practical limits. A country can feel that an expansion is hollow and still drift deeper into it because every local incentive rewards delay.

That is not stupidity. Sometimes it is. But not always. Often it is structure.

The problem is not only that people ignore warning. The problem is also that warning and leverage follow different timelines.

That difference matters. It may be one of the deepest hidden facts of complex systems. A process can become visible before it remains tractable. The signs can arrive while there is still enough uncertainty to delay action, but not enough uncertainty to remove dread. By the time the proof is overwhelming, the options are fewer, uglier, and more expensive. You can end up with the worst combination. Enough information to suffer. Not enough force to steer.

Now move from 1929 to 2008.

The costumes changed. The mechanism did not.

In the years before the global financial crisis, housing was not merely an asset class. It had become a social myth. House prices rose. Lending standards weakened. Credit became easier. Financial engineering multiplied. Risk was bundled, sliced, relabeled, and distributed across the system. On the surface, it looked sophisticated. Underneath, the structure was depending on assumptions that were far less robust than advertised.

Again, the public story was intoxicating. Homes always rise. Diversification spreads risk. Modern finance is smarter now. The institutions are large. The models are advanced. The ratings are reassuring. The system is liquid. The professionals are in charge.

Again, a rising structure produced the illusion of safety.

And again, some people could already see that the arrangement was unstable. They could see leverage stacked on leverage. They could see that a lot of the confidence came from complexity rather than strength. They could see that the whole machine depended on collateral values staying elevated and on funding conditions staying friendly. They could see that one weak link would not stay local because the structure had become too interconnected for local failure to remain local.

But what exactly does it mean to see that.

It does not mean knowing the exact day. It does not mean predicting every detail. It does not mean mystical certainty. It means something humbler and more important. It means detecting the direction of instability. It means understanding that the system is becoming dependent on conditions that cannot be relied upon indefinitely. It means recognizing that what looks like resilience may really be a temporary ability to postpone acknowledgment.

That kind of recognition is real. But it is not the same thing as control.

A trader who sees the danger can still get crushed by timing. A short seller can be right in structure and bankrupt in sequence. A risk manager can understand the exposure and still be ignored. A homeowner can sense the absurdity and still be trapped because rent is rising, prices are rising, and waiting itself is expensive. A regulator can worry and still move too slowly. A citizen can see the madness and still be economically tied to the madness.

Once again, the future became visible before it became touchable.

That is why financial crises are more than economic events. They are demonstrations of a deeper law of human experience. They show us that access to information does not automatically become access to power. They show us that clarity can outlive influence. They show us that a system can confess its danger in fragments while remaining too large, too networked, too incentive driven, and too delayed for clean intervention.

This is not just about finance.

But finance makes the lesson impossible to romanticize.

When a market breaks, there is no place to hide. Losses appear. Liquidity disappears. Names fall. Institutions fail. Retirement accounts shrink. Businesses close. Governments panic. The nice language evaporates. The machinery becomes visible because the machinery is now damaging people in public.

By then, the story is already late.

The truly important interval was before the public confession. The truly important interval was when the system was still being praised, still producing rewards, still covering over its own weaknesses, while those weaknesses were already accumulating. That is the interval FAWP is designed to describe.

Future Access Without Presence is not an excuse for passivity. It is not a way of flattering ourselves after the fact. It is not a mystical halo placed over anxiety. It is a structural concept. It says that in many systems, especially complex systems, predictive contact can survive beyond practical control. You may still be able to extract signal from the present about the future, even after your direct influence on that future has decayed below meaningful force.

If that sounds abstract, markets bring it down to earth.

Suppose the structure is full of leverage. Suppose incentives reward denial. Suppose institutions are entangled. Suppose every major player prefers one more quarter of calm. Suppose ordinary people are financially exposed and emotionally invested. Suppose any individual warning can be dismissed as early, alarmist, or self interested. In that setting, recognition does not spread through the system at the same speed as vulnerability. Vulnerability can build quietly while the official narrative remains smooth. The process becomes visible to some long before it becomes actionable for most.

That is the leverage gap.

The signal arrives before the handle.

You can feel the direction of the machine, but you do not have the torque to turn it.

That image matters because it applies far beyond Wall Street. Once you see it in markets, you begin to notice it elsewhere. A business can already be hollowing out before the layoffs arrive. A currency can already be weakening before the panic becomes public. A family budget can already be unsound before the crisis statement lands. A health decline can already be underway before the formal diagnosis. In all these cases, the structure leaks information before it yields to intervention.

That is why so many people carry a private sentence inside them after something goes wrong.

I knew.

Those two words are often treated as vanity or denial. Sometimes they are. But sometimes they are an honest description of a real condition. Not total knowledge. Not certainty. Directional knowledge. Partial access. Early signal. Enough to feel the coming shape. Not enough to stop the shape.

Financial history is full of those people. The person who knew the leverage was insane. The person who knew the valuations were fiction. The person who knew the lending was rotten. The person who knew the boom depended on a story that could not survive forever. They were not gods. They were not free of error. But they were in contact with reality before reality was officially admitted.

And that contact did not automatically grant them power.

That is the emotional heart of FAWP.

The gap hurts because it violates a story we desperately want to believe. We want to believe that intelligence cashes out into control. We want to believe that good perception is rewarded with leverage. We want to believe that if a danger is visible, someone can surely stop it. Markets teach otherwise. They show that visibility and control are different properties. They can overlap. But they can also separate, and when they separate, the result is a very modern form of suffering.

You see the structure. The structure keeps going.

You read the numbers. The numbers keep smiling.

You understand the instability. The instability is still making people rich.

You know the machine is running hot. The machine is still handing out applause.

That is a hard thing for a civilization to admit, because it offends our faith in information. We live in an age saturated with data, dashboards, analytics, signals, and constant updates. We like to imagine that more measurement means more command. But markets show the limit of that fantasy. You can have measurement without mastery. You can have signal without sufficient force. You can know more and still remain pinned inside a process larger than your reach.

This book is about that condition.

It is about the line where forecasting survives but steering fades.

It is about the painful interval where the future is already leaking into the present as signal, while the pathway from present action to future change has become too weak, too delayed, too noisy, or too entangled to rely on.

And it begins in financial markets because markets teach this lesson with unmatched clarity. They teach that the crash does not begin on the day the newspaper says crash. By then the internal damage has been growing for some time. The public event is late. The structural change came first. The warning came first. The asymmetry came first.

The first real crack in a boom is not always a falling price.

Sometimes it is a hidden dependency.

Sometimes it is leverage that only works while optimism remains cheap.

Sometimes it is a system so interconnected that nobody can tell whether stability is genuine or merely temporary.

Sometimes it is the collective decision to treat delay as safety.

That is the world FAWP tries to map.

Not a world of prophecy.

A world of partial contact.

A world where the future often arrives twice, first as pattern, then as event.

A world where the cruelest knowledge is often not full knowledge, but early knowledge without enough traction.

That is the prologue to this book, and perhaps to much of modern life.

The market is still climbing. The crowd is still smiling. The machine is still paying out. But somewhere beneath the celebration, the structure has already changed.

Some people can feel it.

Almost nobody can stop it.

Chapter 1

There is a feeling almost everyone knows, even if almost nobody has a clean name for it.

It is the feeling of seeing something coming before it arrives, and still not being able to do very much about it.

Not in some magical way. Not in a movie way. Not with thunder in the sky and a glowing prophecy. I mean the ordinary human version, the version that shows up in real life, in tired kitchens, in late night messages, in doctor waiting rooms, in bank balances, in job calls, in a body that suddenly does not feel like your own body anymore. The version that arrives quietly. The version that does not ask permission. The version that can sit beside you for weeks before the world catches up.

You know this feeling if you have ever sensed that a relationship was ending before the words were spoken. Nothing dramatic had happened yet. No final explosion. No official goodbye. But the air had changed. The replies were thinner. The warmth had moved somewhere else. The pauses were longer. The future of that relationship was already leaning in a direction that your mouth could not yet say out loud.

You know this feeling if you have ever felt in your body that something was wrong before any paper or scan or test result proved it. You wake up and know, not with certainty, not like a machine, but with that raw animal knowledge that says this is not normal, this is not me, this is not the same. Then comes the long delay. The appointments. The waiting. The maybe. The we should keep an eye on it. The formal world moves slowly, but your body has already started talking.

You know this feeling if you have ever seen a job going bad before the announcement. Before the layoffs. Before the carefully polished phrases. Before the smiling language that tries to make collapse sound strategic. You can feel it in tone. In timing. In the way people stop promising things. In the way certain names disappear from conversations. In the way the future starts shrinking even while everyone pretends the room is still expanding.

You know this feeling if you have ever looked at money, markets, rent, debt, or your own monthly numbers and understood that a wall was coming long before the wall actually hit. You could already see the math. You could already feel the pressure tightening. But seeing the curve and escaping the curve were not the same thing.

That is the first fact this book wants to take seriously.

Seeing is not the same as steering.

Most people are taught some softer version of the opposite. We are taught that if we are alert enough, smart enough, disciplined enough, informed enough, then we can stay ahead of events. We imagine life as a simple bargain. Notice the danger early, then act. Spot the pattern, then change the outcome. Get the information, then get the power.

Sometimes that is true. Sometimes early awareness really does save you. Sometimes a warning is enough. Sometimes a small action at the right moment changes everything.

But a huge amount of real life does not work like that.

A huge amount of real life works like this instead. You begin to understand what is happening just as your ability to change it starts getting weaker. You begin to see the direction clearly, but the window for easy action is already narrowing. By the time the future becomes emotionally obvious, it may already be harder, costlier, slower, riskier, or messier to move than people like to admit.

That is a brutal kind of clarity.

It is one thing to walk in darkness. At least darkness is honest. It tells you that you do not know. It tells you to be careful. It tells you to move slowly.

It is another thing entirely to see the cliff and realize your brakes are fading under your feet.

That second experience is worse. Not because it contains less information, but because it contains enough information to hurt. Enough information to create dread. Enough information to make you replay choices. Enough information to force one of the ugliest thoughts a human being can have, which is this, I knew. Maybe not perfectly. Maybe not in detail. But I knew enough to feel the shape of it.

That sentence has enormous emotional weight. I knew enough to feel the shape of it.

People carry that sentence through breakups, business failures, health scares, family disasters, market losses, and private collapses of every kind. They carry it with shame because they assume knowing should have given them control. They carry it with self hatred because they assume seeing should have guaranteed action. They carry it with confusion because they cannot explain why awareness arrived before power.

This book begins by saying that maybe there is more going on there than weakness.

Maybe there is structure.

Maybe there are times when the world becomes legible before it remains easily reachable.

Maybe one of the most common painful experiences in modern life is not simple ignorance, and not simple helplessness, but something stranger and sharper. Partial visibility with weak leverage. Recognition without enough traction. Clarity that arrives inside narrowing options.

This matters because the modern world is full of signals. More than ever before, people live surrounded by patterns, alerts, forecasts, feeds, projections, metrics, updates, and warnings. We know more. We measure more. We detect more. We track more. But more visibility has not automatically turned into more human control. In many parts of life, it has produced the opposite feeling. People can sense what is coming, yet still feel out of position to stop it.

That contradiction sits at the heart of this book.

The first step is not to solve it. The first step is to admit it.

There is no point beginning with equations or technical language if the lived reality has not landed first. A real idea should click against life. It should lock onto moments people already know but could not quite explain. That is what I want to do here. I want to begin not with theory, but with recognition.

You have almost certainly felt this in some part of your own life.

You felt it when a person was leaving long before they left.

You felt it when your body knew before the paperwork knew.

You felt it when your finances bent toward trouble before the formal damage arrived.

You felt it when a whole stretch of your life had already started leaning downhill, even while the calendar still pretended everything was normal.

That feeling is real.

It is common.

It is painful.

And it may tell us something profound about the difference between information and influence, between sensing and changing, between seeing and doing.

The world does not always hide its direction from us.

Sometimes it does something crueler.

Sometimes it lets us see the wave before it lets us reach the ocean.

The strange thing is that most people do not talk about this feeling in a serious way.

They talk around it.

They turn it into bad luck. They turn it into nerves. They turn it into overthinking. They turn it into a personal flaw. They say they hesitated. They say they should have been stronger. They say they should have trusted themselves sooner. They say they were stupid. They say they were weak. They say they had the signs right in front of them and still did not move.

Sometimes that self-accusation is fair. Human beings absolutely do avoid things they should face. We delay. We deny. We make excuses. We wait for a cleaner moment that never comes. We hope reality will spare us the cost of acting early. A lot of pain in life really does come from hesitation.

But that is not the whole story.

The whole story is harder, and more interesting.

Sometimes the problem is not that you failed to notice.

Sometimes the problem is that noticing did not give you enough grip.

That is a very different kind of problem. It means the gap between awareness and action is not always moral. It is not always a character issue. It is not always a matter of courage. Sometimes the structure of the situation itself is already working against you. By the time the pattern becomes visible enough to feel real, the road to changing it is already narrower than people want to admit.

Think about how often life works like that.

A person senses that a relationship is cooling. Fine. What is the correct move. Speak immediately, and maybe you sound insecure, suspicious, dramatic. Wait for more certainty, and the emotional distance grows while you collect proof. Push harder for connection, and maybe you create pressure that makes the retreat worse. Stay calm and say nothing, and maybe silence becomes part of the decline. Awareness does not arrive with instructions. It arrives inside ambiguity.

The same thing happens in health. You feel that something is wrong. But feelings are not always evidence in the formal sense. You still need appointments. You still need thresholds. You still need a chain of confirmation. You still need a world outside your own body to catch up with what your body has already begun saying. In the meantime time keeps moving. Maybe nothing serious is happening. Maybe something serious is happening. The point is not that intuition is perfect. The point is that lived recognition and formal traction often run on different clocks.

Money works that way too. A person can see pressure building long before the bill arrives that forces a decision. They can feel the slope turning against them. They know the spending is wrong, or the debt is too heavy, or the runway is too short. But seeing the problem is not the same as instantly having a better job, higher savings, lower costs, or a magical exit. The future can be visible while the practical routes out remain ugly, slow, and expensive.

That is why this feeling hurts in such a specific way. It does not feel like blindness. Blindness is simpler. This feels like partial contact with the truth, mixed with weak leverage. It feels like standing in a room where the fire has not reached you yet, but the smoke already has.

And once you notice how often that happens, a deeper question begins to open.

Why do we assume that seeing and changing should go together in the first place.

That assumption sounds natural, but it hides a very flattering picture of human life. It tells us that if we are perceptive enough, reality will reward us with control. It tells us that awareness is a form of command. It tells us that the person who understands first should also be the person who wins.

That picture is comforting, but the world is often less generous than that.

Plenty of things can remain visible while becoming harder to influence. A market can become more readable just as it becomes more crowded and more dangerous. A relationship can become more emotionally obvious just as the easy repair window is disappearing. A career path can become more clearly doomed just as the available alternatives become fewer. A body can become more alarming just as the path to easy reversal grows smaller.

That is the pattern this book wants to put front and center.

Not ignorance. Not total helplessness. Something in between.

A person knows enough to feel the direction, but not enough, or not early enough, or not forcefully enough, to bend the outcome with confidence.

That middle territory matters because so much human suffering lives there. It is where regret grows. It is where shame grows. It is where people start replaying old scenes in their heads and telling themselves a simpler story than the real one. I knew, therefore I should have been able to stop it. I saw it coming, therefore what happened next must be my fault. I felt the danger, therefore my failure to prevent it proves weakness.

Sometimes. But not always.

Sometimes the painful truth is that a person can be perceptive and still out of position. They can be sensitive and still too late. They can be correct in direction and still unable to convert that correctness into effective action. They can carry real information without carrying enough force.

That possibility changes the emotional climate of the whole problem. It does not erase responsibility, but it does refine it. It allows us to stop treating every failed intervention as proof of stupidity or cowardice. It creates room for a more exact question. Not only did you know. But what could still be done from where you stood, with the time, clarity, cost, and friction available to you then.

That is a harder question, and a much more honest one.

It is also one of the central questions of modern life. Because modern life is full of situations in which people receive more signs than they receive power. More signals than options. More visibility than reach. They can sense how things are moving, but that does not mean they can step outside the current and redirect it.

And that mismatch produces a special kind of loneliness.

Often one person feels the curve first. They live with it first. They carry the future before anyone else is ready to say its name. That can make them sound negative, dramatic, pessimistic, unstable, or obsessed. But sometimes they are simply early. Early enough to feel the shift. Too early to be believed. Too late to feel fully powerful.

That is a brutal place to live. Between silence and proof. Between intuition and permission. Between what you already know in your bones and what the world will only admit later.

So this chapter is not asking the reader to worship instinct, or to romanticize dread, or to pretend that every fear is wise. It is doing something more disciplined than that. It is saying that there is a recurring human experience here, one important enough to deserve a real framework. An experience in which the future becomes legible before it remains easy to touch.

That is the feeling we are beginning with.

Not because it is dramatic. Because it is common.

Not because it is mystical. Because it is ordinary.

And not because it solves anything by itself. Because naming it is the first step toward understanding why awareness and influence so often pull apart.

One reason this feeling has become so common is that modern life is overflowing with signals. You are surrounded by clues all the time.

Your phone tells you how you slept. Your bank app tells you what you spent. Your feed tells you what is trending. Your calendar tells you what is coming. Your inbox tells you what people want. Your body tells you when something is off. Your habits tell you where your energy is going. Your relationships tell you, in tone and rhythm and silence, whether warmth is growing or fading. Your work tells you what direction the floor beneath you is taking. The world is loud with indicators now. It flashes, pings, warns, nudges, predicts, ranks, tracks, and measures.

A person living a century ago could miss a great deal simply because the information never arrived. A person living now can drown in signs and still feel strangely powerless.

That is not a small change. It is one of the defining conditions of the age.

We tend to assume that more visibility should mean more control. The logic sounds obvious. If I can see more, I should be able to respond better. If I can track the pattern, I should be able to get ahead of it. If I know sooner, I should suffer less.

But real life keeps breaking that promise.

A person can track their finances down to the last subscription and still be trapped inside expenses they cannot shrink fast enough. A person can know every calorie, every sleep score, every stress marker, and still watch burnout advancing faster than recovery. A person can sense that a social platform is shaping their mood, attention, and desires, and still find themselves pulled back into the same loops. A person can read endless commentary about the economy, culture, technology, and human behavior, and still feel that all this awareness produces very little direct influence over the forces pressing on daily life.

That is a strange arrangement. More sight, less grip. More warning, less reach. More forecast, not always more power.

In some ways the modern person is like someone standing in a glass room during a storm. The walls are transparent. You can see the weather with incredible clarity. You can watch the clouds gather, the pressure drop, the branches begin to bend. You can tell what kind of night is coming. But seeing the storm through glass is not the same as controlling it. The clarity may even make the helplessness feel worse. You are not confused. You are informed. The trouble is that information, by itself, does not always open a path.

That difference matters because the human mind does not naturally accept it. The mind treats a visible problem as a call to action. It assumes that what can be seen can be touched. That is how we evolved in smaller, simpler environments where the distance between perception and action was often short. Hear the rustle, move your body. See the fire, run. Notice the threat, act.

Modern life is built very differently. It is full of delayed consequences, layered institutions, crowded incentives, hidden chains, and slow feedback. You can detect the movement of very large forces while remaining only a tiny point inside them. That creates an emotional mismatch. Your instincts tell you that seeing should help. Reality answers, not always.

This is one reason people become so exhausted by constant updates. It is not only the volume of information. It is the ratio between information and useful leverage. If the ratio were generous, then awareness would feel empowering. But if awareness keeps arriving without matching reach, then it starts to feel like burden. You end up carrying forecasts you cannot use, warnings you cannot cash in, patterns you cannot cleanly redirect.

Think about how this plays out in ordinary life.

A person reads enough about health to know the broad direction of their habits. They know what stress is doing. They know what poor sleep is doing. They know what chronic overload is doing. But knowledge does not instantly rearrange work demands, family duties, money pressure, emotional history, and the thousand practical forces shaping daily life. The person is informed, yes. They are not imaginary. But they are also entangled.

Or take work. The signs are everywhere. Leadership language shifts. Hiring freezes begin. Projects are quietly delayed. Deadlines get strange. Meetings lose conviction. Everyone can feel the temperature dropping, but nobody is in a position to alter the whole direction. People become spectators to a future that is already entering the room.

Or take relationships. By the time the pattern becomes visible enough to name, that pattern may already have had months to harden. And naming it does not guarantee repair. Sometimes naming it helps. Sometimes it is the only honest thing to do. But the broader point remains. Visibility does not come with a tool kit attached. It comes first as knowledge, and only sometimes as leverage.

This is why so many people now live with a low hum of dread they cannot fully explain. It is not just fear of bad outcomes. It is the repeated experience of seeing enough to worry while lacking enough traction to feel calm. That combination changes the texture of life. It makes people second guess themselves. It makes them wonder whether awareness is useful or just painful. It makes them resent their own sensitivity. It tempts them to shut down, numb out, scroll harder, distract themselves, or retreat into simpler stories.

The simplest story is usually self-blame. If I saw it, why did I not stop it. Why did I not leave sooner. Why did I not prepare better. Why did I not act faster. Those questions can be useful in limited doses, but they become poisonous when they ignore the deeper reality. The modern world often gives you partial contact with the future without giving you clean command over what follows.

That is not a defect in one person. It is built into the arrangement of many parts of life now. Large organizations move slowly. Markets absorb signals quickly. Social attention shifts in crowds. Bodies speak in weak early hints. Platforms learn from you while shaping you. Institutions lag. Emotions harden. Costs rise. Feedback comes late. By the time a pattern becomes obvious enough to feel undeniable, the cheap and graceful options may already be thinning out.

This is where the feeling everyone knows starts to look less like a private weakness and more like a general condition. Not universal, not total, not permanent, but common enough to deserve serious attention. It is the condition of living among growing visibility and uneven leverage. It is the condition of receiving more signs than power.

That may sound bleak, but there is something liberating in saying it plainly. A person can stop pretending that all suffering comes from stupidity. A person can stop assuming that every delayed rescue proves moral failure. A person can start asking a better question.

Not just, what do I know.

But, what can still be moved from here.

That question changes everything. It shifts attention away from the fantasy that awareness automatically creates control. It begins forcing a cleaner separation between two capacities we casually blend together every day. The capacity to see. And the capacity to steer.

That separation is where the next part begins.

There is another reason this experience is so hard to talk about.

It does not arrive as certainty.

It usually arrives as a change in texture.

That matters, because people are very uncomfortable acting on texture. They want proof. They want a chart, a diagnosis, a confession, a number, a clean external sentence that says yes, this is happening, you are now officially allowed to believe what you already felt. Until that moment arrives, they often live in a strange half lit state. They can sense the direction, but they do not yet feel licensed to treat that direction as real.

That gap between sensing and permission is one of the places where Future Access Without Presence grows.

Think about how much of life depends on thresholds. Formal thresholds. Social thresholds. Emotional thresholds. Financial thresholds. Medical thresholds. People often feel a shift long before a process crosses the line required for public acknowledgment. In that period before the line, the future can already be bending. The trouble is that many forms of action only become acceptable after the bend has become obvious to others.

This is one reason late action is so common. It is not always because people are foolish. It is not always because they are cowardly. Sometimes it is because human beings do not live inside pure information. They live inside social timing. They live inside doubt. They live inside other people's standards of proof. They live inside costs. They live inside hope. They live inside embarrassment. They live inside the fear of looking ridiculous if the pattern they feel turns out to be false.

That makes early action expensive.

A person who leaves a relationship at the first deep chill may later wonder if they acted too soon. A person who sounds the alarm about a failing project too early may be seen as negative or unstable. A person who pushes for deeper medical investigation before the formal signs are strong enough may be treated like an anxious exaggerator. A person who cuts risk early in a market can look foolish while the crowd keeps partying for another six months.

This is one of the cruelest parts of the whole experience. Acting early often feels worst exactly when it is most useful.

That is not a poetic idea. It is a practical one. In many areas of life, the cleanest interventions are available before certainty hardens. But before certainty hardens, social and psychological support for action is weaker. So people wait. They wait for stronger evidence. They wait for language. They wait for consensus. They wait for permission. And while they wait, the slope gets steeper.

This does not mean early action is always wise. That would be another fantasy. Early action can be destructive. It can be paranoid. It can create the very damage it fears. The point is not that everyone should obey every uneasy feeling. The point is that uncertainty has a price, and one part of that price is delayed response. When a process is real but not yet fully legible, people become trapped between two risks. The risk of acting too early, and the risk of acting too late.

That is a much harder problem than simple ignorance.

Ignorance is clean. If you do not know, then you do not know. But partial recognition is messy. It gives you enough contact with reality to feel pressure, without giving you enough confidence to move cleanly. It loads the nervous system with anticipation. It makes you watch things more closely. It makes you rehearse conversations that have not happened yet. It makes you scan for confirmation. It makes every new sign feel heavy. You become an unwilling student of a future that has begun arriving before it has fully declared itself.

This is also why people often look irrational from the outside during these periods. They are responding not only to what is formal and visible, but to a growing internal map of direction. To the outside world, nothing major may have happened yet. To the person inside the pattern, many small things have already happened. Tone has shifted. Timing has shifted. Friction has increased. Warmth has drained. Pressure has accumulated. The story has already started leaning.

That difference in access creates conflict.

One person says, you are overreacting.

The other person says, no, you are under reacting.

Neither one is necessarily dishonest. They may simply be standing at different distances from the same process. One has felt the early curvature. The other is waiting for the process to become obvious enough to count.

This is true in intimate life, but it is also true at larger scales. Entire groups can live inside this mismatch. A small number of people may feel that something in a company, a field, a city, an economy, or a culture is changing direction long before broader consensus forms. They carry the emotional burden first. They sound alarmed first. They feel strain first. Then, months or years later, what they felt becomes common language. By then, the interesting question is no longer whether the pattern was real. The interesting question is what kind of leverage still remained once the pattern became undeniable.

That is the heart of it.

When does reality become visible.

When does it become actionable.

Those are not the same moment.

You can already feel the importance of this distinction. If visibility and action always rose together, then life would be simpler. See clearly, move effectively. Learn faster, win earlier. But life often staggers those two moments apart. First comes a weak signal. Then a stronger one. Then a pattern. Then private dread. Then social proof. Then public admission. Then some kind of formal response. By that stage, the world has often already traveled a great distance.

This staggered sequence creates one of the most painful illusions in human life. Looking backward, it can seem as though the outcome was obvious from the beginning. But in the beginning it was not obvious. It was suggestive. It was textured. It was emerging. And acting on an emerging pattern is not the same as acting on a proven one. Hindsight compresses all that ambiguity into a neat story and then uses the neat story as a weapon. You knew. So why did you not act. But real time is not hindsight. Real time is fog with gradients.

That is why this chapter keeps insisting on the lived reality of the feeling. Because without respecting the uncertainty of real time, we misunderstand the whole problem. Future Access Without Presence is not a claim that people possess clean perfect knowledge and somehow fail to use it. It is a claim that people can have meaningful directional contact with what is coming while lacking enough confidence, force, timing, or support to reliably bend the outcome.

That directional contact is enough to change experience.

It changes the atmosphere of a day. It changes how a room feels. It changes how a person interprets silence. It changes how they read small details. It changes the emotional weight of time itself. They are no longer just living in the present. They are living under the pressure of an incoming shape.

That is a heavy way to live.

And the heavier the shape becomes, the more people crave confirmation. Not because confirmation creates the pattern, but because confirmation finally lowers the social cost of believing what was already becoming visible. It says, now you are allowed to trust what you felt. Now you are allowed to act without looking unreasonable. Now the future has become official.

But by then, sometimes, the best opening has already narrowed.

That is why the feeling everyone knows has such a bitter aftertaste. It is not just that things went wrong. It is that people often experience a period in which they were already in contact with the direction of wrongness, without having a clear, low cost, socially supported way to intervene. They were not asleep. They were early in an awkward way. Early enough to feel the pull, late enough to feel the trap.

This is where self blame begins for many people. They look back and confuse early contact with full power. They tell themselves they had the truth in their hands, when what they really had was a growing pattern, a partial map, and a narrowing window. That is still serious. It still matters. But it is not the same thing as full control.

And once you understand that, a different picture of life starts to appear.

The problem is not always that people are blind.

The problem is that reality often becomes legible by degrees, while leverage decays by degrees, and those two curves do not always line up in our favor.

That mismatch, between when you can feel it and when you can still move it cleanly, may be one of the deepest hidden structures in modern life.

And it is exactly the structure we need to understand.

This is the point where the feeling needs a name.

A real name, not a vague mood word, not a slogan, not a little phrase people throw around when life feels hard. A name that says exactly what the experience is, and what it is not.

The name is Future Access Without Presence.

Those three words matter because they keep both halves of the experience alive at the same time. The first half is forecasting. That is the part people are often embarrassed to admit. It is the part where they say, I could feel where this was going. I could sense the direction. I could see the pattern taking shape before the final event arrived. The second half is without power. That is the part people often hide behind shame. It is the part where they say, and still I could not turn that knowledge into enough force, enough timing, enough traction, enough reach, to make the outcome bend the way I wanted.

Put those together and something important happens. The feeling stops being a pile of regrets and starts becoming a recognizable condition.

Future Access Without Presence does not mean perfect knowledge. It does not mean certainty. It does not mean that a person knows every detail of what will happen next. Real life is rarely that clean. Usually the forecast is directional. A person senses the tilt. They feel the pressure building. They notice a pattern hardening. They grasp the broad curve before the destination becomes public fact.

That kind of forecasting is far more common than people admit.

You see it in relationships, where the ending becomes emotionally real long before the final conversation. You see it in health, where the body gives off a low signal that says something is wrong before the formal world speaks clearly. You see it in work, where a role, a team, or a company begins leaning toward trouble before the official language changes tone. You see it in money, where a person can feel the slope of the numbers long before the crisis point is reached. You see it in markets, where direction can become visible even while safe leverage disappears. You see it in culture, where fatigue, distrust, and transition can be sensed before public language catches up.

What people lack in these moments is not always information. What they lack is usable power.

That word power can make people uneasy because it sounds dramatic. It sounds political. It sounds theatrical. But here it means something simpler and more practical. It means the real ability to change the path of events from where you currently stand. Not fantasy power. Not moral power. Not the comforting inner story that says you are still in charge. Real leverage. Real grip. Real reach into what happens next.

That is the painful split. A person may have meaningful contact with what is coming, while lacking enough grip to redirect it in a reliable way.

Once you name that split, a lot of life becomes easier to read.

The person who says, I knew this was coming, may not be flattering themselves after the fact. They may be describing an actual experience of early pattern contact. The reason that sentence sounds so haunted is not because it is always false. It is because it often carries both truth and failure at the same time. Truth about direction, failure of leverage.

That is why people hate the sentence so much.

It is one of the most human sentences in the world, and one of the hardest to live with.

I knew.

Not everything. Not with precision. But enough.

And then comes the second wound.

Why did that not help me more.

Future Access Without Presence is a way of answering that second wound without falling into nonsense. It does not answer by saying your pain was imaginary. It does not answer by saying you had magical access to the future. It does not answer by saying you are weak and useless. It answers by proposing something harder and more exact. It says that there are real situations in life where directional visibility survives longer than practical leverage.

That is a serious claim. It is also a deeply humane one. It makes room for the fact that people can be perceptive without being sovereign. They can be alert without being able to command outcomes. They can care deeply, see clearly enough, and still find themselves standing too far from the machinery that drives the next stage.

This naming also matters because it separates the idea from a few easy confusions.

Future Access Without Presence is not simple helplessness. Helplessness says nothing I do matters. That is too blunt. Future Access Without Presence says something narrower. It says that in some conditions, from some positions, at some moments, your current ability to alter the path is much weaker than your ability to detect the path.

It is not pessimism either. Pessimism is a style of expectation. This is a structure of experience. A cheerful person can live through it. An optimistic person can live through it. A disciplined person can live through it. It is not about personality first. It is about the relation between awareness and leverage.

It is not passivity. A passive person may refuse to act even when clean action is available. That happens. But Future Access Without Presence points to a harsher possibility. Sometimes clean action is already fading by the time the pattern becomes emotionally undeniable. A person may still act. They may act bravely. They may act intelligently. The point is that the world may no longer be offering the same cheap, elegant openings it offered earlier.

It is not prophecy. People love turning painful intuition into mysticism because mysticism is glamorous. But most of the time this experience does not feel glamorous at all. It feels ordinary. It feels practical. It feels like tension, hesitation, math, dread, tone, timing, bodily warning, social lag. It feels like living close enough to a process to notice its direction before the rest of the room is ready to say it aloud.

That is much more interesting than prophecy, because it belongs to daily life.

And daily life is where this book lives.

Once the phrase lands, you begin noticing how many memories reorganize themselves around it. The breakup you once described as a surprise was not really a surprise. It was a slow recognition with a late ending. The work collapse you once called bad luck had a long prehistory of signals you could feel but could not convert into safety fast enough. The personal crisis you once framed as stupidity may have contained months of partial visibility without clean traction. The market loss that felt humiliating may have been a case of direction without usable edge.

Naming does not erase pain, but it changes the shape of pain. It removes some of the fog. It lets a person say, this was not pure blindness, and it was not pure control either. It was something in between. I had contact with the future, but not enough force over it.

That in between zone is where this book begins, because that zone is where so much of modern life is actually lived.

People are taught to think in opposites. Knowing or not knowing. Acting or failing to act. Control or chaos. But real life often unfolds in gradients. Weak signals. Delayed proof. Rising clarity.

Shrinking options. Growing recognition. Thinning reach. The map does not become visible all at once, and power does not vanish all at once either. They move on different clocks.

Future Access Without Presence is the name for what it feels like when those clocks drift apart. That is the point of giving the feeling a title. Not to make it sound clever. To make it discussable. To make it thinkable. To make it possible to ask better questions.

Not just, why did I fail.

But, what exactly did I know, when did I know it, and what real leverage still remained at that point.

Those are better questions. Sharper questions. More honest questions.

They lead away from shame and toward structure. They make life less mystical and more readable. They prepare us for the next step, which is to separate two capacities that everyday language loves to mash together.

The capacity to see what is coming.

And the capacity to still move it.

That split is where the book now has to go, because once you name Future Access Without Presence, you can no longer pretend that forecasting and control are the same thing.

Now that the feeling has a name, the next step is to protect that name from confusion.

That matters more than it seems.

The moment a new idea appears, people rush to flatten it into old categories. They say it is just fear. Or they say it is just intuition. Or they say it is just another word for being passive. Or they say it is fatalism dressed up to sound intelligent. That is what happens whenever a useful distinction enters ordinary life. The mind tries to drag it back into familiar boxes.

So it is worth being careful here.

Future Access Without Presence not the claim that human beings can read the future like a finished script. It is not the claim that every uneasy feeling is a hidden truth. It is not the claim that all bad outcomes are visible in advance. And it is not the claim that action never matters.

It is a narrower and more disciplined idea than that.

It says that there are many moments in life when the direction of events becomes legible before your practical ability to redirect those events remains strong. Not gone in some absolute sense, not erased from reality, but weaker, slower, more expensive, more uncertain, more crowded, more diluted. The pattern becomes easier to see just as your grip on the next stage becomes less clean. That is a very different statement from simple despair.

Despair says there is nothing to do. Future Access Without Presence says there are conditions in which knowing and doing pull apart. Sometimes sharply. Sometimes gradually. Sometimes only for a while. But the split is real enough to shape experience.

That difference matters because despair is dead. It goes nowhere. It collapses everything into one heavy fog. But a split between recognition and leverage can be studied. It can be measured. It can be thought about in a precise way. It can even be improved in some cases, by earlier timing, better feedback, smaller scope, cleaner signals, stronger local action, or lower friction. A person trapped in despair hears only doom. A person who understands the gap between seeing and steering starts asking better questions.

What exactly could still be moved at that stage.

What had already become too costly.

What signs appeared early.

What forms of action were still real, and which were only emotional theater.

That is the kind of thinking this book is trying to build.

Future Access Without Presence is also not laziness. This point matters because modern culture loves moral simplification. If a person saw trouble coming and failed to stop it, the easiest explanation is that they were weak, soft, undisciplined, avoidant, or foolish. Sometimes that explanation is partly true. People do hide from reality. People do wait too long. People do talk themselves out of difficult choices. There is no need to romanticize human hesitation.

But the moral story becomes stupid when it assumes that awareness automatically comes with clean leverage. Real life is messier than that. A person can be alert and still trapped inside delayed consequences. A person can be disciplined and still too small relative to the force moving around them. A person can be brave and still lack the right timing. A person can act and still discover that the action channel is weaker than they thought.

That is why the phrase matters. It does not flatter people. It clarifies the terrain they were standing on.

It says that failure can come from many places. Some of those places are personal. Some are structural. Some come from fear. Some come from bad timing. Some come from a weak signal. Some come from a narrowing window. Some come from being one small actor inside a process that had already gathered more momentum than one local move could easily reverse.

Once you admit that, the emotional weather around many memories begins to change.

A person may look back on a breakup and realize that what haunted them was not the fantasy of total control, but the knowledge that they had been in contact with the direction of the ending long before they had a clean way to stop it. A person may look back on a financial collapse and realize that what felt like stupidity was partly a long period of visibility without a workable exit. A person may look back on burnout, illness, family strain, or career decline and realize that their deepest pain came from knowing enough to worry, without knowing enough, or moving early enough, or possessing enough real force, to bend the path.

That is not self absolution. It is more interesting than self absolution. It is a better map.

Future Access Without Presence is not pessimism either. Pessimism is a stance. It is a style of expectation. Some people lean dark by temperament. They anticipate trouble. They expect disappointment. They brace early. But this idea is not about mood. It is about the relationship between information and influence. A cheerful person can live through it. A hopeful person can live through it. A highly motivated person can live through it. The gap does not care what your personality test says. It belongs to the structure of the situation.

This is important because it keeps the book anchored in reality. The point is not that some special tragic personality type notices doom before others. The point is that the world often reveals direction before it preserves easy reach. That can happen to anyone, in love, in health, in money, in work, in public life, in private habits, in every corner of ordinary existence.

Future Access Without Presence is not magical intuition. That one is tempting, because people love turning pain into mystique. If someone repeatedly senses the direction of events before others do, it is flattering to imagine they possess a special gift. But most of the time the truth is less glamorous and more useful. Human beings pick up weak signals all the time. Tone shifts. Rhythm shifts. Incentives shift. Friction increases. Numbers stop behaving the way they used to. Rooms feel different. Bodies feel different. None of this requires supernatural language. It requires attention, pattern sensitivity, and life inside processes that are already changing shape.

That is what makes the phenomenon so ordinary, and so important. It does not belong only to mystics, analysts, traders, doctors, or philosophers. It belongs to anyone who has ever felt the future leaning before the world around them was ready to say so.

And yet the idea does not belong to intuition alone either. That matters too. We should not romanticize gut feelings. Human beings overread noise constantly. We scare ourselves. We project. We catastrophize. We invent patterns where none exist. We feel danger because we are tired, ashamed, lonely, traumatized, or overwhelmed. The existence of Future Access Without Presence does not mean every warning inside the mind deserves obedience.

What it means is that the experience itself deserves better language than we usually give it.

Instead of saying either I am just anxious, or I have a mystical gift, a person can start saying something much more useful. I may be in contact with a real directional pattern, but the question is whether I still have enough practical leverage to alter it, and how much confidence the signal actually deserves.

That sentence is not poetic. It is operational. It pushes the mind away from shame and fantasy and toward structure. It asks for discrimination. It asks for timing. It asks for signal quality. It asks for honest assessment of what can still be touched from here.

And that is the deeper value of the concept. It teaches restraint in two directions at once. It stops people from dismissing weak but real pattern contact as mere irrationality. And it stops people from inflating every uneasy feeling into sacred prophecy. It builds a middle ground. A serious one.

This middle ground is where a lot of adult life actually happens. Not in certainty. Not in blindness. In gradients. In partial maps. In rising evidence. In narrowing options. In signals that grow clearer while leverage grows worse. In warnings that matter, but not enough on their own. In situations where truth arrives by degrees and action loses strength by degrees, and those two curves do not line up in the kind way we wish they would.

That is why this book is not beginning with formulas. It is beginning with recognition and then with discipline. The recognition says, yes, this feeling is real. The discipline says, now let us keep it clean.

Not superstition.

Not self-pity.

Not moral laziness.

Not doom worship.

A real pattern, in ordinary life, where visibility can outlast easy control.

Once that is clear, we are finally ready to make the next move. We are ready to separate, in a more exact way, the power to forecast from the power to steer. Because until those two capacities are split apart, the feeling everyone knows remains emotionally vivid but conceptually blurry.

The blur ends next.

So this is where this leaves us.

Not with a solution yet. Not with a theorem. Not with a machine that explains everything in one clean movement. It leaves us with recognition, which is the first thing any serious idea has to earn. Before a concept deserves trust, it has to touch reality. It has to meet people where they already live. It has to name something that was there before the name arrived.

That is what this chapter has been trying to do.

It has been trying to show that one of the most common painful experiences in ordinary life is not simple ignorance. It is not pure chaos. It is not always weakness, or passivity, or fantasy. Very often it is something much stranger and much more precise. A person begins to feel the direction of events before they possess enough clean leverage to change that direction in a reliable way. They are not fully blind, but they are not fully in command either. They stand in a narrow strip of reality where recognition is rising and power is thinning.

That strip is where a great deal of modern life is lived.

It is lived in marriages that are still intact on paper and already broken in tone. It is lived in bodies that have started whispering before the formal world speaks loudly enough to count. It is lived in work that still looks stable from far away and already feels hollow from the inside. It is lived in money that has not yet become catastrophe and has already become pressure. It is lived in friendships, families, careers, cities, markets, habits, and private lives of every kind.

What makes this experience so corrosive is not simply that things go wrong. Things have always gone wrong. What makes it corrosive is that wrongness often arrives with warning, but warning does not always arrive with reach. The future can begin pressing on the present before the present offers a graceful way to answer back.

That is a hard truth, but it is not a hopeless one.

In fact, one of the quiet hopes inside this whole book is that a more honest map can reduce some unnecessary suffering. Not all suffering, because life does not make that bargain. But some. A person who keeps interpreting every failed intervention as a total failure of character will slowly poison their own perception. They will stop trusting what they notice. They will confuse delayed leverage with personal stupidity. They will begin to believe that seeing is worthless because seeing did not save them in the form they wanted.

That would be a mistake.

Seeing still matters.

It matters even when it does not grant total command.

It matters because reality does not become more merciful when ignored. It matters because weak signals, even when imperfect, are often the first contact we have with what is changing. It matters because early awareness can still shape smaller decisions, softer landings, earlier honesty, reduced exposure, less self deception, less theater, less wasted time. It may not always let a person stop the wave, but it can sometimes help them stop pretending the sea is still flat.

That is not nothing.

In some ways it is the beginning of maturity. Not the fantasy of being all powerful, but the discipline of asking what is visible, what is still reachable, and what has already begun to harden beyond easy control. Those are better questions than the moral slogans people usually use on themselves. They are sharper, cleaner, and more humane.

That is why this first chapter had to stay close to feeling. Not because feeling is enough, but because feeling is where the theory enters life. If the reader does not recognize the experience, then the formal concepts later will sound decorative. But if the reader does recognize it, if they can feel in their own memory the strange pressure of seeing enough to hurt without seeing enough to command, then the rest of the book stops being abstract. It becomes an explanation of something already lived.

That is where we are now.

We have named the condition. Future Access Without Presence.

We have separated it from several easy distortions. It is not simple despair. It is not prophecy. It is not a permission slip for passivity. It is not merely a bad mood. It is not every anxious thought. It is a more disciplined claim. There are conditions in which directional visibility survives longer than practical leverage. There are situations in which people can meaningfully detect where things are going while lacking enough timely influence to redirect the path in a strong or reliable way.

Once that possibility is admitted, the mind changes shape around it.

Past memories change shape around it too.

Old failures begin looking different. Not always smaller, but clearer. The breakup that once looked like total surprise becomes a story of long recognition and weak traction. The health fear becomes a story of early bodily contact and late formal confirmation. The career collapse becomes a story of visible drift and narrowing options. The financial blow becomes a story of clear slope and inadequate room to maneuver. The emotional tone of these memories shifts because the underlying structure becomes more legible.

That matters because clarity is not only for science. It is also for mercy.

A good concept should make the world more exact, but it should also make experience less stupid. It should stop us from using blunt language where finer language is needed. It should help us stop calling everything weakness when some of it is timing. It should help us stop calling everything irrationality when some of it is genuine contact with an emerging pattern. It should help us stop calling everything fate when some of it is simply late leverage.

This is the deeper promise of the book.

Not to flatter people.

Not to tell them they secretly knew everything all along.

Not to dissolve responsibility into a soft fog of excuses.

The promise is more serious than that. The promise is to separate what we usually mash together. To separate awareness from control. To separate signal from reach. To separate forecasting from steering.

That separation is the real gateway.

Because the moment you admit that these are different capacities, a new set of questions appears. What does it mean, exactly, to forecast. What counts as usable information about what comes next. What does it mean to steer. What counts as real influence over what comes next. Can one remain alive while the other weakens. Can they be measured separately. Can a person, or a system, cross a boundary where action becomes too delayed, too weak, too noisy, or too diluted to matter, even while prediction still survives.

Those questions take us into the next step.

That is where the book now has to go. Out of recognition and into distinction. Out of the felt life of the problem and into its first clean conceptual split. Because until forecasting and control are pried apart, people will keep blaming themselves with the wrong vocabulary. They will keep assuming that if they saw it, they should have been able to move it. They will keep confusing knowledge with leverage.

It begins by breaking that confusion.

Prediction is not control.

That is the hinge.

That is the split.

That is where the science begins.

Chapter 2

Most people grow up with a quiet assumption about life. If you can see clearly enough, then you should be able to act clearly enough. If you understand what is happening, then you should be able to change what happens next. Knowledge, in that old picture, feels like the first half of control. First you notice, then you move. First you learn, then you steer. First the mind catches up, then the world bends.

It is a beautiful picture. It is also incomplete.

One of the deepest mistakes people make, in love, in health, in money, in public life, in private habit, is treating prediction and control as if they were the same power wearing different clothes. They are not. They can overlap. They can help one another. In some moments they sit very close together. But they are not identical, and once life becomes large, delayed, noisy, crowded, or tangled, the gap between them begins to widen.

That gap is where so much confusion begins.

A person says, I knew this was coming, so why could I not stop it. A family says, we all saw the signs, so why did nothing change. A whole culture can feel the drift of something before it finds a way to interrupt that drift. Again and again people punish themselves for a hidden belief they rarely say out loud. If I could see it, then I should have been able to steer it.

This part begins by breaking that belief.

To see where something is going is one kind of power. To still be able to alter where it is going is another. The first concerns information. The second concerns leverage. The first asks what the present can tell you about what comes next. The second asks whether what you do now still has enough force, enough timing, and enough clean contact with the future to make the future change shape.

Those are different questions.

Think about weather. Humanity has become remarkably good at forecasting large patterns in the sky. We can watch pressure move across continents. We can model wind, heat, moisture, and storm formation. We can often tell, days ahead of time, whether a hurricane will likely curve north or stay low, whether a heat wave will intensify, whether a cold front will arrive with violence or pass with only rain. That is prediction. It is real, impressive, and useful. Yet no one mistakes a weather forecast for control over weather. No one says that because the storm is visible, the storm is obedient. The forecast can be excellent while the power to redirect the storm remains basically absent.

That sounds obvious in meteorology. It sounds less obvious when the same logic enters ordinary life.

Take the human body. Sometimes the body starts giving off signals before a condition becomes formally clear. A person feels wrong in a way that is difficult to explain, but impossible to ignore. Sleep is broken. Energy falls through the floor. Appetite shifts. Pain appears without permission. Mood becomes oddly physical. The body is saying something, even if it is not speaking in full sentences. That is a kind of prediction. The present contains clues about what may be unfolding. But acting on those clues does not always mean the future can still be turned easily. Appointments take time. Tests take time. Interpretation takes time. Some processes can be visible before they remain easy to reverse. Again, information and leverage do not arrive together like perfect twins.

Now think about driving. Imagine a car whose steering responds with a long delay. You turn the wheel now, but the tires answer much later. In such a car, a person may see danger perfectly well. They may understand the curve, the speed, the road, the threat, all of it. Their prediction may be excellent. But prediction is not enough if the channel between action and outcome has become slow and unreliable. The road does not care that the driver understands the problem. What matters is whether the action can still reach the future in time. A driver with delayed steering can be informed and endangered at the same time.

This simple image carries enormous weight. It tells us something that many people resist because it is emotionally offensive. Reality does not always reward awareness with command. Sometimes you can read the curve and still miss the turn, not because your mind failed, but because the connection between present action and future outcome became too degraded.

In small and simple situations, the difference can hide. Touch a hot stove, pull your hand away. Hear a sound in the dark, turn your head. Watch a glass tip near the edge of a table, catch it. In these close, local, low delay moments, prediction and control can feel almost fused. Seeing and moving seem like one smooth act. This is one reason human intuition so often assumes they belong together. In much of immediate daily motion, they do.

But once scale grows, once distance grows, once delay grows, once many moving parts interact, the old intuition begins to fail.

A market can become readable while remaining hard to trade safely. A relationship can become emotionally legible while remaining difficult to repair. A population can see a problem and still find no clean way to reverse it. A person can know a habit is dragging them toward misery and still find that knowing does not instantly produce enough force to break the pattern. In each case, prediction survives in one form while control weakens in another.

That is not a contradiction. It is a structure.

A useful way to feel the distinction is this. Prediction asks whether the future is giving off enough shape to be noticed from here. Control asks whether your hand, your action, your choice, your intervention, still reaches far enough into that future to reshape it. One is about visibility. The other is about grip. One is about legibility. The other is about reach. They can travel together, but they do not have to.

This should change the emotional meaning of a great many memories. People often replay the past as if foresight and influence were one thing, then punish themselves because one survived while the other did not. They remember sensing the end of a relationship and conclude that they should have been able to prevent the end. They remember seeing pressure build in their finances and conclude that seeing the slope should have been enough to escape the slope. They remember sensing exhaustion, drift, fragility, dishonesty, dependence, and assume that noticing the pattern should have granted mastery over it. But that assumption smuggles in a false law of the universe. It assumes that if a future is visible, it is still easily touchable. Often it is not.

This is why so many people live with a bruised relationship to their own perception. They felt the curve of events, yet their feeling did not save them in the form they wanted. So they begin to doubt the value of seeing at all. They confuse failed control with false perception. They start telling themselves that awareness is useless, or worse, that awareness is just another way to suffer. The real lesson is harsher and more exact. Awareness matters, but it does not guarantee leverage. A forecast can be meaningful long after the action channel has started to decay.

That sentence sits at the heart of everything that follows.

A forecast can be meaningful long after the action channel has started to decay.

Once that becomes thinkable, a whole new map opens. You can begin to ask sharper questions. Not only, what did I know. But also, what kind of contact did my action still have with the future at that point. Not only, did I see it coming. But also, was the road between present choice and future outcome still short, strong, and clean enough to matter. Not only, was the pattern visible. But also, was it still pliable.

This is the first great split the book needs to make. Without it, everything stays morally muddy. People keep blaming themselves with the wrong language. They keep assuming that information and influence rise and fall together. They do not.

Seeing and steering are separate powers.

Prediction and control are separate capacities.

And modern life, with all its scale, delay, friction, and complexity, often lets one survive while the other begins to thin.

That is the door we have now opened.

A useful way to understand this split is to stop thinking about knowledge as one thing.

We often talk as if knowing were a single solid block. Either you know or you do not. Either you understand or you do not. Either you saw it coming or you did not. But the world is more subtle than that. There are different kinds of knowing, and they do different jobs.

One kind of knowing tells you what is likely to happen next. It reads direction. It detects shape. It sees where the curve is leaning. This is the kind of knowing that lets a sailor read the sky, a doctor read a pattern in symptoms, a trader read crowd pressure, or a tired person read the drift of their own life. It is about the future becoming legible from clues that already exist in the present.

Another kind of knowing is not really knowing at all, at least not in the ordinary sense. It is practical reach. It is the ability to put a hand into the stream of events and alter the stream while it is still close enough, soft enough, and responsive enough to be altered. This is not about reading the direction of things. It is about touching the direction of things.

Those two capacities often travel together in simple moments, so close together that people confuse them. A child sees a ball rolling toward the edge of a table and reaches out to catch it. Awareness and intervention feel like one act. A cook sees a pot about to boil over and lowers the heat. A driver sees a bicycle swerve and taps the brake. In moments like these, the road from noticing to doing is short. The feedback is quick. The world answers promptly. Cause and effect still hold hands.

This is one reason people develop such a deep emotional faith in the idea that clear sight should produce clear control. In many local moments it does. The mind learns from thousands of tiny victories. See the thing, move the hand, change the outcome. Notice the danger, respond, survive. Understand the pattern, act, improve the result. Those lessons become part of our emotional architecture. We do not merely believe them with the mind. We feel them in the body.

Then life grows larger.

Distance enters. Delay enters. Noise enters. Many moving parts begin to interact. Consequences arrive slowly, or in layers, or through other people, or through forces too large for one local action to redirect neatly. The old faith remains, but the world begins to violate it.

Think again about weather, because weather is a clean teacher. A meteorologist may know quite a lot about a storm. They may understand the path, the pressure, the likely intensity, the timing of landfall, the vulnerability of a region, even the places where damage is most likely to concentrate. That is real knowledge. It can save lives. It can guide preparation. It can change choices. But it is still not command over the clouds. No forecast, however elegant, reaches upward and rearranges the atmosphere. The future in that case is visible, but not obedient.

Now move from storms in the sky to storms in ordinary life.

Imagine a doctor watching the arc of an illness. Sometimes the pattern is visible before intervention becomes easy. The signs are there. The trajectory can be read. The likely future is already leaning in one direction. But knowing the trajectory is not the same as possessing unlimited power over the body. Treatments have timing. Bodies have thresholds. Human beings arrive late, delay care, misread symptoms, lack resources, fear results, or simply live inside the slow pace of formal confirmation. The doctor may understand what is likely. The understanding itself does not erase friction.

Or imagine a person watching their finances. They can read the slope of their own situation. They can tell that expenses are outrunning income, that the margin is too thin, that one shock could trigger a much larger collapse. That is a form of genuine insight. But insight does not instantly produce savings, better wages, lower rent, lower debt, or a second life hidden behind a curtain waiting to step onto the stage. The future can be visible while the route out remains narrow, painful, and uncertain.

Or imagine a relationship in slow decline. One person may feel the direction first. They sense it in the rhythm of replies, in the shrinking generosity, in the way laughter no longer lands the same, in the way ordinary closeness begins to require effort. That is real contact with the future of the relationship. But there is no law saying that such contact automatically grants the means to reverse the drift. Speak too early and you may sound unstable. Speak too late and the emotional distance may already be deep. Push for clarity and you may trigger defense. Stay silent and the pattern keeps hardening. Awareness and leverage are not the same gift.

This is the point where many readers begin to feel a quiet release. Not because pain disappears, but because the map becomes more honest. A person can stop assuming that every failed intervention proves stupidity. A person can stop treating every painful forecast as though it should have come packaged with a steering wheel. The world does not make that bargain.

A simple machine can help make this visible. Think of a thermostat. A thermostat is very good at control within a narrow range. It does not meditate on the future. It does not imagine tomorrow. It does not feel dread. It does not narrate. It simply measures the current temperature and adjusts the room toward a target. In that narrow local setting, it has strong practical reach. It changes what happens next in a direct and measurable way.

Now compare that with a weather forecast. The weather forecast may carry much richer knowledge about what is coming, but it possesses no comparable reach into the storm itself. One thing has low insight and strong local control. The other has high predictive power and almost no direct steering. Put those side by side and the old confusion begins to crack. Seeing and steering are different capacities.

The same lesson appears in driving. A car with responsive steering lets understanding become action quickly. The wheel turns, the tires answer, the road changes. But now imagine the wheel is delayed, or loose, or badly connected. The driver may be alert, skilled, and fully aware of the coming curve. None of that guarantees safety if the connection between present input and future movement has become weak. Here again, prediction survives while control starts to decay. The driver knows. The road does not care. What matters is not only whether the mind saw the curve, but whether the hand still reaches the future in time.

That line matters.

What matters is whether the hand still reaches the future in time.

So much human suffering comes from forgetting that phrase. People judge themselves as though reality were always a close responsive room, a place where insight naturally translates into influence. But many of the situations that matter most are not close responsive rooms. They are stretched across time. They are filtered through other people. They are burdened by fear, cost, delay, pride, denial, complexity, and scale. They may be legible without being pliable.

That is why prediction can feel like both a gift and a wound. It lets you see the contour of what is coming. It gives shape to fog. It lets you orient yourself. But if the action channel has already weakened, prediction may also deliver the strange ache of clarity without rescue. You are no longer blind, but you are not yet free.

This is not a reason to devalue prediction. Quite the opposite. Prediction matters enormously. Weather forecasts save lives even if they cannot calm the sky. Medical insight matters even if it cannot guarantee reversal. Emotional clarity matters even if it cannot restore a fading bond. Financial realism matters even if it cannot instantly erase pressure. Forecasting can still guide preparation, honesty, reduction of harm, smaller wiser moves, earlier truth, and cleaner decisions. The point is not that forecasting is useless. The point is that forecasting is not the same as control.

Once that lands, a new question begins to rise.

If these are truly separate capacities, then what determines whether the second one remains alive. What determines whether an action taken now still has enough strength, enough speed, and enough contact with the future to matter. At what point does the road between present choice and later outcome grow too weak, too delayed, or too noisy for reliable influence.

That is where the deeper structure begins.

For now, one thing is enough.

The future may be visible from here.

That does not mean your hand still reaches far enough to reshape it.

Why do people confuse these two powers so easily.

Why does the mind keep acting as if seeing and steering belong together.

Part of the answer is ancient. Human beings were not built to spend most of their lives dealing with giant, delayed, crowded arrangements. We were shaped in a world of near things, local things, immediate things. A sound in the grass. A branch underfoot. Fire too close to the hand. A face changing expression across the space of a room. In that kind of world, awareness and action often live close together. You notice, you move. You read the situation, you answer it. The world replies quickly enough that your body learns a simple lesson. Seeing matters because seeing changes what happens next.

This lesson sinks deep.

It becomes more than a belief. It becomes a reflex. It becomes part of our emotional logic. That is why people feel insulted by the idea that knowledge may not produce control. It offends a very old instinct. If I can tell what is happening, then surely I should be able to do something meaningful about it. If I cannot, then something must be wrong with me.

That conclusion feels natural. It is also often false.

The older world of local response still exists. It exists every time you catch a falling glass, step away from danger, answer a child calling your name, reach for a coat when rain begins, lower the flame under a pot, or tap the brake at the right second. In moments like those, awareness and influence still travel together. The mind learns again and again that the world is responsive, that contact is immediate, that the future is near enough to touch.

Then a person steps into adult life, and the scale changes.

Now the important things are stretched out over time. They are tangled with money, institutions, habits, emotion, distance, delay, crowd behavior, social pressure, and long chains of cause and effect. You can still see a great deal. In some ways you can see more than ever. But the road from seeing to steering is no longer short and straight. It winds. It narrows. It gets muddy. It runs through other people, through inertia, through uncertainty, through time itself.

This is where the old instinct begins betraying us.

A person senses burnout long before a collapse. But the body is not a light switch. Rest competes with duty, fear, debt, ambition, guilt, identity, routine, and the ordinary stubbornness of life. The future is visible in one sense. The path to altering it is crowded.

A person sees a relationship cooling. But there is no guaranteed move that restores tenderness by decree. Love is not a steering wheel with instant response. The future of the bond may already be leaning while every available action carries risk. Speak, wait, soften, confront, retreat, ask, listen, pause, press. Each move may change something, but not in the clean way the heart wants. Again, the future is visible before it is easily manageable.

A person knows their work is becoming unstable. The signs are there in tone, timing, shrinking confidence, drifting priorities, and the quiet nervousness that spreads through a place before official words admit anything. But seeing the drift does not instantly produce a better offer, a safer landing, or an easy exit. The hand reaches out, and the world answers slowly.

This is why modern life creates so much silent self-accusation. People are still using an ancient emotional rule inside a much larger arena. They are asking a slow crowded world to behave like a nearby object. They are asking large delayed processes to respond like a falling cup. When that does not happen, they blame their own weakness instead of noticing the altered terrain.

The altered terrain matters.

A close object can be stopped by a hand. A long process often cannot.

A near danger can be dodged in an instant. A social pattern, a financial slide, a medical condition, a crowded market, a cultural shift, or a long private habit is different. It unfolds through layers. It resists simple correction. It may grow more visible over time while also becoming more stubborn.

That is the part people hate.

They do not merely hate bad outcomes. They hate the possibility that outcomes can become clearer while becoming harder to alter.

They hate the idea that awareness may arrive as witness before it arrives as force. They hate the fact that truth can reach them before power does.

Yet this is one of the most ordinary facts of life.

Think about a large ship at sea. If you stand on the deck and look far enough out, you may be able to see a hazard long before you reach it. That is a gift. It gives you time. But a ship is not a bicycle. It does not pivot on a fingertip. Even perfect recognition does not mean instant maneuverability. Mass matters. Momentum matters. Turning radius matters. Delay matters. The ocean does not care how early the lookout saw the danger if the vessel cannot change course quickly enough.

Human lives contain many versions of that same problem.

People imagine that seeing earlier should feel like safety. Sometimes it does. Sometimes early sight is the difference between damage and escape. But early sight can also create a different burden. It can create the burden of carrying a future before other people are ready to acknowledge it. It can create the burden of watching momentum continue while your available leverage feels weak, uncertain, or socially unsupported. It can create the burden of knowing enough to worry but not enough to command.

This is why knowledge can feel so lonely.

The person who senses the change first often suffers first. They are not living in a different world, exactly. They are living in the same world on a different clock. They have already felt the bend in the road while others still see only straight pavement. They have already heard the distant thunder while others are still discussing whether the sky looks all that strange. Their perception moves ahead of shared permission.

This gap between private recognition and public action is one of the great engines of frustration.

A person may know they need to leave a job, leave a relationship, cut a habit, reduce risk, slow down, change direction, tell the truth, seek help, or prepare for impact. But knowing is not the same as moving. Motion asks for timing, courage, resources, cooperation, and often sacrifice. The world rarely grants all of those at once. So the person lives inside an ugly in between. They have enough contact with the future to feel strain, but not enough traction to feel relief.

That in between can last months or years.

It can happen in love, where the heart knows before the conversation happens. It can happen in health, where the body knows before the file is thick enough to trigger decisive movement. It can happen in finance, where the slope is obvious before the runway is gone. It can happen in public life, where a crowd feels the drift long before the drift becomes reversible.

And because the mind is still carrying its old local instincts, it interprets this in between as failure. If I can see it, why can I not turn it. If I understand it, why does my understanding not carry more force. If I am aware, why does awareness not buy me peace.

The answer is not that awareness is worthless.

The answer is that awareness and leverage belong to different layers of reality.

One layer concerns legibility. Is the future giving off enough shape to be noticed from here.

Another concerns reach. Does present action still have enough contact with what comes next to bend it.

In simple local moments, those layers sit close together. In larger delayed situations, they drift apart. A person may have one without possessing much of the other. That is not a personal defect. It is a feature of the terrain.

Once this becomes clear, a more intelligent form of self-judgment becomes possible. Instead of asking only whether you knew, you can ask whether the road from action to outcome was still strong. Instead of saying I should have been able to stop it because I saw it, you can ask what kind of leverage truly remained at that point. Instead of worshipping awareness as if it were command, you can begin to see it for what it often is, an important but incomplete form of contact.

That is a major shift.

It softens nothing false. It excuses nothing cheap. But it clears away a great deal of unnecessary shame. It lets a person admit that they were not blind, while also admitting that the world they were trying to steer had already become large, delayed, and resistant.

And once that admission is made, a harder and more beautiful question begins to rise.

If knowledge and leverage are separate, what governs the second.

What decides whether a present action still reaches the future with enough strength to matter.

What decides whether the road between now and next is short and clear, or long and fading.

That is where this path is going.

If there is one force that quietly pulls prediction away from control more than any other, it is delay. Delay is one of the least dramatic forces in life, which is exactly why people underestimate it. It does not arrive with flames or alarms. It does not look heroic or evil. It looks ordinary. A pause. A lag. A wait. A few days. A few weeks. A conversation postponed. A result not yet back. A feeling that sits in the body before language catches up. A pattern that becomes visible before the world around it is ready to move.

Yet delay has enormous power. It can turn insight into frustration. It can turn warning into memory. It can take a clear signal and make it operationally useless. In many important parts of life, a person does not fail because they had no information. They fail because the information arrived into a world where the distance between now and next had already become too long, too crowded, or too slippery for action to travel cleanly.

This is one of the hardest ideas for people to accept, because delay insults common sense. Common sense says that if you know earlier, you should be better off. And in many local situations that is true. If you see a stone rolling toward a glass, one second matters. If you notice a bicycle swerving toward your lane, one second matters. If a child reaches toward fire, one second matters. In short response situations, the road between seeing and doing is short enough that a little early warning can change everything.

But many of the things that shape a life are not short response situations. They are stretched across time. They develop through layers. They harden while people are still discussing them. They gather force while signals are still being debated. In these cases, knowing early in one sense may still be too late in another. The person has a forecast, but the forecast is arriving into a process that has already built momentum.

Think about illness. There are times when the body begins speaking long before formal confirmation arrives. A person feels wrong. Not imaginary wrong, not theatrical wrong, but wrong in a way that quietly reorganizes the day. Stairs feel different. Sleep feels different. Hunger feels different. Recovery feels different. Something has changed. Yet the path from that feeling to decisive action is rarely instant. There is doubt. There is scheduling. There is cost. There is waiting. There is the ordinary hope that things will settle on their own. There is the slow machinery of tests, referrals, interpretation, and thresholds. During all that time, the body is not paused. The process continues moving. Delay stretches between recognition and response, and every stretch of delay changes what can still be done easily.

The same thing happens in relationships, though people often resist describing it so plainly. Emotional distance usually does not fall from the sky in one afternoon. It accumulates. A person notices the cooling, then doubts themselves. They notice it again, then decide not to overreact. They wait for a better moment. They hope the mood will lift. They tell themselves they are tired, stressed, dramatic, unfair. Meanwhile time continues doing its quiet work. Habits of tenderness weaken. Defensive stories take root. Small disappointments pile up. By the time the problem becomes clear enough to feel undeniable, the easy version of repair may already be gone. This does not mean nothing can be done. It means delay has already changed the cost of doing it.

Money offers another clean lesson. A person can often see financial pressure long before the formal crisis. The numbers are there. The pattern is there. The runway is shrinking. The margin is too thin. One more shock would hit much harder than it should. Yet seeing that curve does not instantly create a new income, new savings, lower rent, or a safer structure. The warning is real, but the world answers slowly. A person may know exactly what direction things are heading while still lacking enough immediate leverage to change the direction before the consequences arrive with full force.

This is why delay is not just a practical nuisance. It is a structural divider between prediction and control. It separates the moment when the future becomes visible from the moment when action remains strong. And the farther apart those moments drift, the more painful life becomes.

There is something almost cruel about this. Human beings tend to imagine danger as a point, a line you cross, a single bad day when everything becomes obvious. But much of life is shaped by gradients. Pressure builds by degrees. Drift builds by degrees. Noise builds by degrees. And because the change is gradual, people often imagine they still have more time than they really do. Delay hides inside gradual change. It whispers that tomorrow is close enough, that next week will do, that another month of watching cannot hurt.

Sometimes that whisper is harmless. Often it is not.

A person who acts on day one may look premature. A person who acts on day ninety may look wise. But by day ninety the terrain may no longer be the same terrain. Delay does not merely postpone action, it alters the field on which action lands. A body may be harder to heal. A bond may be harder to rebuild. A market may be harder to exit safely. A habit may be harder to break. A crowd may be harder to turn. A life may be harder to redirect. Every day of lag changes the geometry of influence.

This is why it is possible to know early in a psychological sense, and still know late in an operational sense.

That distinction matters enormously.

Psychological early means the person has felt the direction before the outcome becomes official. They can sense the bend in the road. They know something is wrong, or off, or leaning, before others admit it.

Operationally early is different. It means an action taken at that moment still has enough clean contact with the future to change the future at acceptable cost.

Those are not the same thing.

A person can be psychologically early and operationally late. In fact, that may be one of the most common painful experiences in adult life. They sense the future before the group does, before the institution does, before the paperwork does, before the public story does. But because the relevant process is delayed, entangled, expensive, or already carrying momentum, their insight arrives without matching leverage. They know first, but they cannot turn first.

This should immediately change how we think about regret. Regret often tells a very simplified story. It says, I knew, therefore I should have acted, therefore what followed proves weakness. But delay complicates that story. The more stretched the process, the more dangerous that simplification becomes. A person may indeed have needed to act sooner. But sooner may still not have been simple, obvious, cheap, socially supported, or clearly effective. Delay makes the moral picture less theatrical and more exact. It forces us to ask not only whether the person saw the pattern, but whether the road from seeing to shaping remained open enough at that stage.

This is also why modern life produces so much exhausted awareness. Large structures move slowly in some ways and suddenly in others. News travels instantly, but coordination lags. Signals flood in, but action routes remain narrow. A person can be rich in warnings and poor in leverage. They can understand the general direction of a process while feeling that every available move is partial, compromised, late, or weak. The result is a nervous system full of anticipation and not enough release. The mind sees the weather coming, but the hand cannot easily move the sky.

This is not a reason to disrespect prediction. Forecasting still matters. It lets people prepare, simplify, reduce exposure, speak earlier, refuse comforting lies, and stop mistaking silence for safety. But once delay enters, forecasting and steering can no longer be treated as one thing. The warning may arrive long before the world is ready to answer it well.

That is the hard lesson.

Delay can preserve visibility while draining control.

The future may become clearer and clearer, while the channel through which you might still change it grows thinner and thinner.

Once you understand that, a great many situations stop looking mysterious. They start looking tragic in a more disciplined way. Not because people knew nothing. Not because they had no care. But because time itself had already begun widening the distance between awareness and reach. Delay is only one reason prediction and control can split apart. Another is that action does not simply exist or fail. It does not behave like a light switch that is either fully on or fully off.

In real life, action weakens. It loses force. It gets blurred, scattered, slowed, bent, misread, or absorbed before it reaches the place where it was supposed to matter.

This is one of the most important facts people miss when they think about control.

They imagine control in heroic terms. Either I can change this, or I cannot. Either my choice matters, or it does not. Either I am steering, or I am just a passenger. But most of life happens in the middle. Your influence may still exist, yet arrive weaker than you hoped. It may still travel forward, yet lose shape as it moves. It may still matter, but not cleanly, not cheaply, not predictably, not enough to produce the result your mind had pictured.

This weakening can happen in many ways.

A signal can be delayed. A response can be noisy. An intention can be misunderstood. A message can pass through too many hands. A small input can be swallowed by a large system. A good decision can arrive into a bad environment. A clear intervention can be diluted by the fact that other people are also acting, resisting, drifting, denying, or simply moving in different directions. By the time your present choice reaches the future, it may no longer look like the same choice.

That is what makes control so much stranger than people want it to be.

Think about trying to steer a boat with loose cables. The wheel turns, but not all the turning reaches the rudder. Some motion is lost in slack. Some force disappears in friction. Some response arrives late. The person holding the wheel is not doing nothing. The boat is not ignoring them completely. But the connection between input and outcome is no longer crisp. It has become mushy. Soft. Unreliable. The difference between a clean response and a sloppy one can mean everything.

A lot of human life feels like that.

You can say the right thing to someone you love, and the words still arrive bent by history, stress, pride, fear, old resentment, exhaustion, and timing. The action was real. The intention was real. But the path between what you meant and what landed was crowded. Some of your force got lost on the way.

You can make a financially responsible choice, and still discover that the wider conditions around you are so harsh that the choice buys far less relief than expected. The action was not fake. It mattered. But it entered a field already loaded with rent, debt, inflation, uncertainty, dependency, and time pressure. Your move reached the future, but not with the strength you imagined.

You can take care of your health, and still learn that care itself moves through layers. Sleep helps, but stress remains. Exercise helps, but timing matters. Treatment helps, but only if it arrives while a process is still soft enough to answer well. A body is not a simple machine with one button and one output. It is a living field of coupled processes. The action channel exists, but it can be noisy, partial, and full of lag.

Once you see this, a more exact picture of influence begins to appear. Influence is not only about whether you act. It is about what happens to the action after you act.

Does it travel quickly.

Does it stay coherent.

Does it get drowned by background noise.

Does it arrive at the right scale.

Does it meet a system that is still responsive.

Does it face competing forces much larger than itself.

Those questions matter because they tell you whether present action still has a strong line into what comes next, or whether that line has begun to fray.

Noise is one of the biggest reasons action weakens. In ordinary language, noise means all the extra disturbance that makes a signal harder to separate from the clutter around it. In a room full of shouting people, one calm voice may be wise but hard to hear. In a market full of violent movement, one good trade idea may be correct but difficult to execute cleanly. In a stressed household, one honest conversation may matter, but it may land inside such a thick atmosphere of fatigue and defensiveness that its effect becomes difficult to detect. The action is there, but the environment is too crowded for the action to stand out sharply.

This is why people often overestimate the power of isolated intention. Intention matters, but intention is not a magic beam that cuts through any level of noise. A signal that is strong in a quiet room can become weak in a loud one. The exact same action can have very different consequences depending on what surrounds it.

Then there is dilution. Dilution is what happens when your action is only one small contribution inside a much larger flow. In a simple local situation, your move may dominate the outcome. You catch the glass, and the glass does not fall. But in larger systems, your action joins many other actions, pressures, incentives, delays, and reactions. It does not vanish, but it becomes one ingredient among many.

This is why social and economic life can feel so frustrating. A person can act responsibly and still feel carried by forces far beyond their own scale. A group can see a problem and still fail to reverse it because no one action has enough weight on its own. A whole public can be full of warnings while still moving along a path that seems increasingly obvious and increasingly difficult to redirect. The problem is not always that nobody acted. Sometimes the problem is that action entered a river too wide for one hand to turn.

Feedback matters here as well. If you act and the world answers clearly, you can learn. You can adjust. You can refine. But if the feedback is poor, delayed, misleading, or hidden, then even a good action can become hard to sustain. You do not know whether it worked, whether to repeat it, whether to stop, whether the system is responding at all, or whether some other force is masking the result. In that kind of environment, control becomes clumsy. Not because the person is foolish, but because the channel between effort and learning has become blurred.

A gardener learns quickly when a plant is thirsty. Water it, and the answer soon appears. But many parts of adult life do not answer that way. The consequences of a decision may take months to show themselves. The meaning of a conversation may unfold long after the words. The effect of stress may emerge slowly. The danger in a financial pattern may only become obvious when the room for graceful adjustment is already smaller. Weak feedback makes control harder because it breaks the rhythm between act, observe, adjust.

This is one reason large systems can feel almost haunted. Not because they are mysterious in some supernatural sense, but because they do not answer human action at human speed. You do something, and the effect disappears into layers. Was it too weak, too late, too small, too scattered, too noisy, too isolated. Or is it still working its way forward. In a system like that, prediction may remain easier than control. You can see direction in the large pattern even while struggling to prove that your small present move still has strong enough contact with the next stage.

This should also change how we think about courage. People love praising decisive action, and sometimes that praise is deserved. But courage alone does not remove friction. Courage alone does not cancel noise. Courage alone does not stop dilution. A brave action can still enter a world that is slow to answer it, crowded against it, or too weakly connected to let it travel far. This does not make courage meaningless. It makes courage only one ingredient in a larger structure.

What matters is not just whether you move, but what kind of road your movement has to travel. A clean narrow road lets even a small action matter a great deal.

A long muddy crowded road may eat most of that action before it reaches the future.

This is why people so often feel the difference between being right and being effective. They are not the same. A person can read the pattern correctly, choose sensibly, even act honorably, and still find that their influence arrived too diluted to reshape events in the way they hoped. The world does not always answer present action with equal clarity. Sometimes it gives back a firm response. Sometimes it gives back fog.

Once that becomes visible, the split between prediction and control becomes even sharper. It is not only that the future may be visible before it is easy to change. It is that the channel of change itself may be thinning, fraying, and losing coherence long before it disappears entirely. Influence does not usually vanish in one dramatic instant. It weakens. It degrades. It gets absorbed into larger motion.

That is the deeper lesson here.

Control is not only about intention.

It is about transmission.

It is about whether what you do now still reaches what happens next with enough force, enough clarity, and enough fidelity to matter.

And once that question enters the room, another truth begins to appear.

A system may still be telling you what is likely, even while no longer giving you enough grip to do much about it. Once delay, noise, and dilution have entered the picture, a person can find themselves in one of the strangest conditions in life. They are no longer blind, yet they are not truly effective. They have information, but not enough leverage. They can read the direction, but their present reach into that direction has become weak, expensive, uncertain, or badly timed. This is where the split between forecasting and control becomes emotionally undeniable.

It helps to say this very plainly.

Information is not the same thing as leverage.

Those two things often travel together in simple situations, which is why people confuse them. If you see a glass sliding toward the edge of a table, the information and the leverage arrive almost at once. You notice the danger, and your hand can still reach the outcome before the outcome hardens. In that kind of moment, awareness feels powerful because the world is still close. The future is only a second away, and your body is still allowed inside that second.

But many of the most important parts of life do not work that way.

A person may know their marriage is in trouble, yet not know how to touch the next stage of that trouble without making it worse, or without paying a cost they are not yet ready to pay. A person may know their finances are drifting toward danger, yet still lack the income, the timing, the margin, or the clean escape route needed to alter that drift quickly. A person may know that a habit is eating their life, yet still lack the force to break the loop before the loop tightens again. A whole population may feel that something in the culture is bending in an ugly direction while still lacking a simple local move that can reverse the bend. In each case the information is meaningful. The leverage is weak.

That distinction changes everything.

It changes how we interpret regret. It changes how we interpret self blame. It changes how we interpret clarity itself. People often act as though the value of information should be measured by whether it produced the exact result they wanted. If the warning did not save them in the most satisfying way, they begin to treat the warning as worthless. But that is far too crude. Information can matter even when leverage is poor. It can matter because it lets a person prepare, simplify, reduce exposure, speak honestly, stop pretending, or choose the less catastrophic path. These are not small things. They are often the difference between injury and ruin.

Think again about a storm at sea. A captain may not be able to erase the weather, but a good forecast still matters. It changes route planning. It changes speed. It changes what gets secured, what gets abandoned, what risks become unacceptable, what timing becomes urgent. The forecast does not command the ocean, but it still has value because it changes how a human being meets the ocean.

This is a very important shift in thinking. Once people stop treating leverage as the only kind of usefulness, they begin seeing the world more clearly. Forecasting may not always let you stop the wave, but it can still tell you not to build your house where the wave is about to land. It can still tell you that denial is no longer cheap. It can still tell you which lies have become too expensive to keep.

That is why it is such a mistake to think that information without control is meaningless. It is not meaningless. It is painful, yes. It is limited, yes. But it is not empty. It may still be the difference between being crushed by surprise and being wounded with some degree of preparation. Human beings often underestimate the value of preparation because they confuse preparation with surrender. They imagine that if they cannot fully reverse a process, then any smaller adjustment is pathetic. That is pride speaking, not wisdom.

Real life is often shaped by partial leverage.

A person may not be able to save a failing bond, but they may still be able to leave it with more honesty and less humiliation. A person may not be able to erase a health problem, but they may still be able to respond earlier, ask better questions, reduce secondary damage, and stop wasting precious time on false comfort. A person may not be able to stop a market collapse, but they may still be able to reduce exposure, stop chasing fantasy, and choose survival over ego. A population may not be able to reverse a large public drift overnight, but individuals within it may still be able to see more clearly, prepare more intelligently, and stop mistaking visibility for safety.

In all these cases the forecast has value even when the steering wheel has grown loose.

Still, none of this cancels the ache at the center of the experience. The ache comes from knowing that the world is giving you a warning while also limiting what the warning can buy. It is one thing to be lost in the dark. It is another thing to stand under a bright light with weak hands. That is why this split hurts so much. It offends our sense of fairness. If reality is willing to reveal its direction, why is it not equally willing to remain pliable. Why let the pattern become visible if the channel of change is already shrinking.

The answer is that visibility and pliability belong to different layers of reality. A process can become easier to read because the pattern is hardening. That is one of the darkest little truths in life. Sometimes the reason a future becomes clearer is precisely that it has gathered so much structure that its path is becoming less flexible. What feels like better understanding may partly be the result of the thing itself becoming more committed to its own direction.

That is why late clarity can feel so bitter.

A relationship is easier to understand once the damage has become regular. A financial slope is easier to see once the margin has grown thin enough to stop lying. A career decline is easier to read once the tone has shifted across the whole field. A public drift is easier to name once enough signs have piled up that denial starts looking theatrical. Clarity rises, but it may rise together with hardness. The pattern becomes more legible because it has become more real, more organized, more committed, more costly to alter.

This should instantly make sense of many human experiences. People often say that in hindsight the signs were obvious. Of course they were obvious later. By later, the pattern had become denser. It had repeated itself enough times to stand out. It had accumulated enough evidence to look solid. But the same growth in legibility may also have meant that the easy moves were already gone. The process had not only become clearer, it had also become stiffer.

That is the cruel trade. Better visibility can arrive together with worse leverage.

Once this is understood, a great many moral confusions start to dissolve. A person no longer has to treat every forecast as a promise of full rescue. A person no longer has to despise themselves because seeing did not immediately become command. Instead they can begin asking more exact questions. What kind of value does this information still have. What smaller adjustments remain possible. What reduction of harm remains possible. What preparation remains possible. What truth telling remains possible. What fantasies need to be dropped right now because they are stealing time that no longer exists.

This is not a small upgrade. It is a huge one. It moves a person from wounded pride toward operational honesty. That phrase matters here. Operational honesty. It means facing not the future you wish existed, but the future your present leverage can still reach. It means refusing to pretend that all options remain equally alive just because you do not want to grieve the narrowing of the field. It means respecting the difference between what is visible and what is still movable.

This is why information without leverage is not just a philosophical curiosity. It is an everyday condition. It appears whenever a human being can see enough to know that the world is moving, while also feeling that their own move will not travel cleanly enough to reshape that movement in the way they want. It appears in health, in money, in love, in work, in public life, in private struggle. It is one of the signature feelings of being alive in a large delayed world.

And once you see it, another idea begins to emerge, almost by force.

If forecasting and control can separate, and if leverage can decay while information remains alive, then perhaps there is some boundary beyond which action becomes too weak, too delayed, too noisy, or too diluted to matter reliably. Not a magical line painted on the sky. Not a dramatic point where reality suddenly snaps from possible to impossible. Something more subtle and more useful. A point, or a zone, where practical reach falls below what a human being can count on.

You can already feel that idea forming.

There must be a point where the road between now and next grows too long or too damaged for dependable steering.

There must be a point where the future is still giving off shape, yet your hand is no longer arriving with enough force.

There must be a point where awareness survives after control has begun to fail.

The next step is to name that boundary, and to see what it means for freedom, for timing, and for the strange pain of living in a world that often reveals its direction before it leaves us enough power to redirect the path.

By now the split should feel less like an abstract idea and more like a fact about lived experience.

A person can understand what is unfolding and still be unable to redirect it with confidence. A future can become legible while practical reach is already thinning. Information can remain alive while leverage becomes weak, delayed, expensive, or scattered. Once this is seen clearly, a great many moments in life stop looking like private madness and start looking like encounters with structure.

That change in perspective matters.

It matters because most people carry a brutal moral interpretation of their own hardest memories. They look backward and tell themselves a story built on one false equation. I saw it, therefore I should have controlled it. I knew enough to worry, therefore I should have known enough to prevent it. I felt the direction, therefore my failure to stop the outcome proves weakness. This equation is simple, flattering, and often wrong.

The truth is more demanding than that.

To see is not to steer. To detect is not to command. To know where a process is leaning is not the same as still having a strong hand on the process. These are different capacities, and one of the deepest confusions in life comes from acting as though they rise and fall together.

Sometimes they do rise together. In a small room, in a local moment, in a clean and responsive setting, awareness and influence can still move almost as one. You notice the danger, you act, the world answers. You see the spill, you stop the spill. You hear the change in an engine, you pull over. You catch the glass before it breaks. In those near conditions the future is close enough to touch.

But many of the forces that shape a life do not stay close enough to touch.

They gather across time. They pass through other people. They grow inside habit, culture, money, emotion, delay, and momentum. They become visible through repetition, through pressure, through a slow thickening of pattern. By the time they are easy to name, they may already be harder to bend. This is why so many forms of clarity arrive with grief attached. The pattern is clearer because it has had time to become stronger.

That is a painful sentence, but it explains a lot.

The pattern is clearer because it has had time to become stronger.

A bond becomes easier to read once distance has settled into routine. A financial slope becomes easier to read once the margin has become too thin to hide behind comfort. A personal habit becomes easier to read once the cost has become visible in mood, body, money, time, or self respect. A public drift becomes easier to name once the signs are no longer isolated. Better visibility does not always mean better power. Sometimes it means the process has hardened enough to reveal its own direction.

This is one reason late clarity can feel so humiliating. A person imagines that because the truth is obvious now, it should have been equally usable before. But obvious now and usable before are not the same thing. The earlier version of the truth may have been weak, partial, costly, socially unsupported, or hard to act on without looking foolish. Later, when the signs have piled up, the person judges their earlier self with the comfort of accumulated evidence. This is one of the ways hindsight cheats.

Hindsight speaks with the confidence of a future that has already happened. Real time never gets that luxury. Real time lives inside uncertainty, price, hesitation, imperfect signal, and incomplete permission. Real time asks a person to move without the safety of total confirmation. Real time forces action to compete with doubt. Once that is remembered, many memories become easier to read honestly. Not softer, not cleaner, but more exact.

This exactness is one of the real gifts of separating forecasting from control. It gives a person a better language for failure, and a better language for success as well. Success is not only getting the exact outcome you wanted. Sometimes success is recognizing that leverage is shrinking and changing your stance before the full impact arrives. Sometimes success is reducing exposure. Sometimes it is telling the truth early. Sometimes it is preparing the ground. Sometimes it is refusing one more month of denial. Sometimes it is stepping out of fantasy while there is still enough road left to walk with dignity.

That is why forecasting still matters, even when control is weak.

A forecast may not let you erase the weather, but it can still change how you meet the weather. It can tell you not to keep building in a flood path. It can tell you that the open sea is no place for pride. It can tell you which stories in your life have become too expensive to keep repeating. It can tell you that a future you hoped was still soft has started becoming hard. It can tell you that delay is no longer neutral. It can tell you that waiting is now a choice with teeth.

None of this turns knowledge into a magic key. That fantasy has to go. The universe does not owe a person control just because it has offered them a glimpse of direction. But neither does the absence of total control make awareness trivial. Once the world is seen through this lens, the task changes. The task is no longer to demand that every warning come with a steering wheel. The task is to ask, with discipline, what kind of contact still remains between now and next.

This is the more mature question.

Not only, what do I think is coming.

But also, what can still be influenced from here.

That question does not flatter the ego, which is one reason people avoid it. The ego prefers fantasy control or complete defeat. It prefers saying I can fix everything, or nothing matters. Both extremes are emotionally simpler than reality. Reality usually offers something messier. Some things remain open. Some things are closing. Some actions still travel forward with force. Some now arrive too late. Some truths are visible but not yet shared. Some changes are possible but costly. Some windows are wide. Some are narrow. Some are gone and only dignity remains.

To live well inside that landscape, a person has to stop worshipping awareness as if awareness were command. They also have to stop dismissing awareness because it did not grant command. The real work is finer than that. It is learning to judge the quality of the road between present action and future result. A short clear road allows a small move to matter greatly. A long delayed noisy road may swallow even a strong move. Most of adult life is spent somewhere between those extremes.

This is why so much depends on timing.

Timing is not just a practical detail that sits at the edge of the story. It is near the center of the story. A choice made while the road is still short may feel small and ordinary. The same choice made later may feel dramatic, costly, and weak. A truth spoken early may guide. The same truth spoken late may only describe damage that has already gathered force. A risk reduced early may feel almost boring. The same risk reduced late may feel like panic. The event may look similar from the outside, but the geometry of influence has changed underneath it.

Once that is understood, another idea comes into view almost by necessity.

If prediction and control are separate, and if delay, noise, friction, and dilution can widen the gap between them, then there must be a boundary of some kind. Not a theatrical line. Not a supernatural wall. Something more practical than that. A region where action becomes too delayed, too weak, too crowded, or too poorly connected to matter reliably, even while the future still gives off enough shape to be read.

People know that boundary in ordinary life even when they do not yet have language for it. They know the feeling of a moment after which the easy move is gone. They know the sadness of realizing that what was once pliable is now stiff. They know the pressure of watching the future become clearer while the door into that future becomes narrower. They know what it means to feel that the hand is no longer arriving in time.

That boundary is the next step.

It has a name.

The Agency Horizon.

Beyond it, action may still exist in some sense, but not with the clean dependable force people imagine when they talk about control. Before it, leverage is still alive enough to trust more directly. Around it, life becomes strange. The future is visible. The road from now to next is degrading. Timing becomes fate in slow motion.

That is where this path now leads.

Into the boundary itself, into the question of how practical reach fades, how late becomes too late, and how freedom depends less on grand slogans than on whether your hand still reaches the future while the future is still soft enough to answer.

Chapter 3

A great many disasters do not begin as disasters.

They begin as small things that still answer back.

A little pain in a tooth. A faint noise under the hood. A credit card balance that feels irritating, but not yet frightening. A conversation that goes cold in a way that could still be warmed. A body that starts sending out weak signals, poor sleep, constant fatigue, rising tension, the feeling that ordinary effort suddenly costs more than it should.

In the beginning, these things still feel manageable. That is what makes them so dangerous. They do not arrive wearing the clothes of catastrophe. They arrive dressed as inconvenience. They arrive in forms that invite delay. A person looks at them and thinks, not today, I can handle this later, this is still small, this is still under control.

And sometimes that is true.

But often it is only true for a while.

One of the most important facts about ordinary life is that many problems do not simply get bigger over time. They become less responsive. They become more expensive. They become harder to interrupt. The same kind of action that would have worked earlier starts working less cleanly later. And that change, from responsive to stubborn, from cheap to costly, from awkward to entrenched, is one of the central hidden structures of adult life.

Take something simple, like a cavity.

At first it is small. It may not hurt at all. Or it hurts only a little, only when the drink is too cold, only when you chew on the wrong side, only at night. It is still a warning, but a warning with a gentle voice. At that stage the problem is not pleasant, but it is still close. A dentist can deal with it quickly. The cost is annoying, but not ruinous. The pain is real, but not yet commanding the whole nervous system.

Then time passes.

The tooth does not stay where it was while you are postponing the appointment. The cavity deepens. The nerve gets closer. Inflammation builds. One day the problem that was once a filling becomes a root canal, or an extraction, or an infection so fierce that it reorganizes your whole week around pain. The same mouth, the same person, the same basic problem, but no longer the same terrain. What changed was not just the size of the damage. What changed was the responsiveness of the situation. Early, the problem answered quickly and cheaply. Later, it answered slowly, brutally, and at much higher cost.

The same thing happens in a car.

At first there is a sound. A soft knock. A slight slip. A tremor that appears only under certain conditions. Something is off, but the car still runs. You can still get to work. The radio still works. The engine still turns over. The problem is not yet dramatic enough to force action. So people wait. They tell themselves they will deal with it after the holiday, after payday, after the next bill, after life calms down.

But machines do not wait with us.

A worn belt becomes a breakdown. A fluid leak becomes a seized part. A small repair becomes a dead transmission, a tow truck, and a bill that feels almost personal in its cruelty. Again, the point is not just that the problem got worse. The point is that the same action no longer buys the same result. Earlier, a modest repair restores function. Later, the road back is far more expensive and far less graceful.

Debt works this way too, and perhaps even more ruthlessly, because debt adds time to the problem. A person looks at a balance and thinks it is manageable. Not good, but manageable. A few payments, a little discipline, maybe next month will be better. But interest does not share human optimism. Interest keeps moving. A balance that once looked irritating becomes a structure. Minimum payments stop reducing the principal in any meaningful way. The person is still paying, still trying, still technically acting, but the debt has crossed into a new regime. The same effort now buys less relief. Time has thickened the problem. A thing that could have been corrected early now behaves like a trap.

There is something almost scientific in the cruelty of that shift.

At first, a small push changes the outcome.

Later, the same push barely moves anything.

That is not just true in money and machines. It is true in human bonds.

A relationship rarely breaks in a single instant. More often it cools. Tiny resentments go unspoken. Small disappointments are absorbed, then remembered, then quietly counted. Tone changes before declarations change. Replies lose warmth before the idea of leaving becomes speakable. In the early stage, the problem may still be soft. A good conversation may still have real power. A sincere apology may still clear the air. A moment of honesty may still prevent months of distance.

Then time passes.

The words not said start accumulating mass. Misunderstanding hardens into interpretation. Interpretation hardens into identity. One person becomes the cold one, the other becomes the difficult one, the exhausted one, the needy one, the unreachable one. Once a bond reaches that stage, the same kind of action that might once have restored closeness can now feel too small, too late, or strangely artificial. The issue is no longer one moment. It has become a climate.

That matters.

Many people think a worsening problem is just a larger version of the same problem. But often it is not. Often it has changed category. A late stage problem is not simply a bigger early stage problem. It is a different kind of thing. It has more momentum.

More resistance. More cost. More history attached to it. It answers differently.

The human body teaches the same lesson over and over.

Poor sleep does not usually destroy a person in one week. Chronic stress does not always announce itself with one dramatic collapse. Weight gain, blood pressure, fatigue, brain fog, low mood, constant inflammation, these often arrive by increments. At first, the body is still negotiating. It complains, but softly. A person can still imagine that a little rest, a little discipline, a better month, a good weekend, will reset everything.

Sometimes it will.

But often the body is already moving from signal into pattern. Fatigue starts changing behavior. Behavior starts changing metabolism. Stress starts changing appetite. Poor sleep starts changing mood, focus, and judgment. Once enough of these loops reinforce one another, the person is no longer dealing with one small problem. They are dealing with a web. Earlier, the body was asking. Later, it is demanding.

This is why so many adults live inside the same painful sentence.

I knew something was off.

That sentence is usually true, but incomplete. The deeper truth is this. The problem was once more responsive than it is now. There was a period when the road from noticing to fixing was shorter. The issue had not yet gathered enough structure to resist correction. That does not mean the action would have been pleasant. It does not mean certainty existed. It does not mean the right move was obvious in the way it looks obvious later. It means the situation was still soft enough to answer.

That softness is one of the most valuable things in life, and one of the easiest things to waste.

Because softness does not feel urgent.

A cavity that does not yet hurt feels optional. A car noise that appears only sometimes feels postponable. A relationship problem that has not yet exploded feels deniable. A balance that can still be paid next month feels survivable. A body that still gets through the day feels good enough.

This is why so much of adult life is not about spotting catastrophe. It is about respecting small trouble before it hardens into larger trouble. It is about understanding that many problems live in windows. Early in the window, the world still answers gently. Later in the window, it answers with greater force and less mercy. After a certain point, the same kind of intervention no longer produces the same kind of result.

That is the key idea this first movement has to establish.

Not all problems stay equally fixable over time.

Some are responsive early, stubborn later, and brutal after that.

Some ask first, then warn, then punish.

Some begin as friction, then become structure.

Once a reader really feels that, everything that follows becomes clearer. The issue is not only whether a person notices a problem. The issue is whether the problem still lives in a stage where action can reach it cheaply and effectively. A weak signal in a soft system is one thing. A clear signal in a hardened system is another.

And that difference may shape more of ordinary life than most people ever stop to realize.

If small problems can grow into expensive problems, then the next question is simple.

What exactly are we losing when that happens.

The usual answer is control. People say they lost control of their finances, control of their health, control of the situation, control of the relationship, control of the direction of their life. That language is understandable, but it is still too loose. It sounds dramatic, yet it does not tell us much. It turns a real process into a foggy feeling. If this book is going to be useful, then we need something sharper than that.

We need a more practical meaning of agency.

Most people hear that word and think of philosophy. They think of free will, destiny, the soul, human choice in the grand cosmic sense. They imagine old arguments about whether people are truly free or whether everything is predetermined. Those arguments can be interesting, but they do not help much when your roof is leaking, your debt is snowballing, your body is wearing down, or your marriage is drifting into silence. In ordinary life, people do not need a grand answer first. They need a usable one.

Here, agency means something very simple.

It means your present action still has usable contact with what happens next.

That is all.

If you do something now, and that action still reaches the future in time, with enough strength and enough clarity to make a reliable difference, then agency is alive. If that contact becomes weak, delayed, scattered, or too costly, then agency begins to fade. The issue is not whether a person technically can act in some abstract sense. The issue is whether acting from here still works well enough to count on.

That definition matters because it clears away a lot of confusion.

A person may still have choices and still have weak agency. They can choose, but their choices no longer travel forward with much force. They may still have freedom in the legal or philosophical sense, yet little practical reach into the outcome they care about. They may still be making decisions every day, yet those decisions are now arriving into a situation that has become stiff, crowded, delayed, or expensive to move.

Think of a steering wheel in a car.

When the steering system is healthy, a small turn of the wheel changes the direction of the vehicle almost immediately. The road is still close. The response is quick. The future of the car, where it will be two seconds from now, is still tightly linked to what the driver does in the present. That is a very clean form of agency. The action and the outcome remain in close contact.

Now imagine the steering becomes loose, delayed, or damaged. The driver can still turn the wheel. The act itself still exists. But the meaning of the act has changed. The road between present motion and future direction has become unreliable. A turn that should have produced a clean result now produces a weak one, or a late one, or a messy one. The driver still has a wheel in their hands, but their reach into what happens next is fading.

This is much closer to how many real life problems work.

Take money. If a person catches a spending problem early, small changes may still matter a lot. Cancel a few things, cut a few habits, adjust the budget, build a little space, and the future responds. The link between present effort and future relief is still fairly strong. But later, if the person is buried under compounding interest, late fees, rising costs, and months of accumulated strain, the same category of action no longer buys the same result. Cutting a few expenses may still be wise, but it does not restore the old flexibility. Agency is weaker not because the person became less real, but because the connection between effort and outcome has degraded.

The same is true in health.

A person who starts walking more, sleeping better, eating more carefully, and reducing stress while the body is still responsive may feel a real difference. The future answers back. Energy improves. Weight shifts. Blood pressure improves. Mood steadies. The body is still in a conversation with the person. But later, when the same body has spent years inside poor sleep, chronic overload, inflammation, hormonal disruption, and reinforcing habits, present action may still matter, but it no longer lands with the same clean force. Progress becomes slower. The system is harder to move. The body is not refusing action completely, but it is no longer as responsive to modest intervention. The distance between effort and result has grown.

That is a loss of agency in the practical sense.

Not because the person is powerless in a theatrical way. Not because all action is useless. But because action no longer has the same quality of contact with what comes next.

This definition also helps with relationships, where people often confuse emotion with power. A person may care deeply, understand the problem, want the bond to survive, and still have weak practical agency. They can talk, apologize, explain, reassure, reach out, try again. But if trust has worn thin, if resentment has built structure, if both people are now reading each other through weeks or months of disappointment, then present action no longer travels through clear air. It passes through memory, suspicion, fatigue, and interpretation. The action is real. The care is real. But the line from now to next has become crowded. The bond is no longer easy to steer.

That is why agency is such a useful word once it is stripped of its grand costume.

It lets us ask a better question.

Not, do I still have choices.

But, do my choices still reach the future with enough strength to matter reliably.

That is a much more honest question.

It also explains why people can feel trapped even when they technically still have options. A person may still be able to leave the job, call the bank, book the appointment, tell the truth, change the habit, have the conversation, sell the asset, ask for help. In that sense, choice remains. But if all those moves now come with higher cost, weaker effect, slower feedback, and less dependable return, then agency has already changed. The person is not frozen. They are operating in a field where action has become harder to convert into results.

This is the difference between formal choice and practical reach.

Formal choice means you can still do something.

Practical reach means that what you do still lands with force.

People suffer because they confuse the two. They tell themselves that because they still technically can act, they should still feel fully powerful. But that does not follow. A person can still act from a position of degraded agency. They can still move from a place where the road between action and outcome has grown long, rough, and expensive.

This becomes even clearer if we think in terms of time. Agency is not just about whether an action exists. It is about whether the action arrives soon enough to matter. A roof leak fixed early is one problem. A roof fixed after weeks of water damage is another. A hard conversation before the emotional freeze sets in is one thing. The same conversation after distance has become habit is another. A spending change before interest compounds is one thing. The same spending change after debt has hardened is another.

In all these cases, action remains possible. But action is no longer standing in the same relationship to the future.

That is why agency is best understood as contact.

Close contact, quick contact, useful contact.

The kind of contact where a present move still reaches what comes next while what comes next is still soft enough to answer.

Once that image lands, a lot of life becomes easier to read. You can stop asking whether a person was simply strong or weak, wise or foolish, brave or passive. Those questions are often too blunt. The more useful question is whether the person still had good contact with the future they were trying to change.

Sometimes they did.

Sometimes they did not.

Sometimes what failed was not effort, but contact.

That is the idea that has to settle here before we go any further. Because once agency is defined as usable contact with outcomes, the next question becomes unavoidable.

What damages that contact.

What stretches it, weakens it, delays it, and makes a future that is still visible begin to slip out of practical reach.

If agency means usable contact with what happens next, then one force damages that contact more quietly than almost any other.

Delay.

Delay does not look dramatic. It does not enter the room shouting. It does not feel like collapse when it first appears. It feels like tomorrow. It feels like next week. It feels like one more month, one more paycheck, one more conversation later, one more visit after things calm down, one more season before I deal with this properly. Delay wears the face of reason. That is why it is so powerful.

Most people imagine danger as a loud event. A crash. A breakup. A diagnosis. A missed payment. A resignation. A public failure. But in ordinary life the real damage is often done much earlier, in the soft stretch of time when the problem is already there, but still quiet enough to negotiate with. That stretch is where delay does its work. It takes a situation that still answers gently, and slowly moves it into a state where the same actions buy less, cost more, and arrive into a harder world.

Think about paperwork.

Not glamorous paperwork, just the ordinary kind that rules modern life. A tax letter you do not open today. A form for insurance that sits on a counter. A visa renewal. A contract that needs a signature. A missed payment notice. At first, these are not disasters. They are tasks. Slightly annoying, slightly boring, easy to postpone because nothing seems to explode when you ignore them for forty eight hours. But time in these systems is not neutral. Late fees appear. Deadlines close. Accounts freeze. Penalties multiply. What could have been handled with ten minutes and a clear head later demands hours, calls, explanations, stress, and sometimes money you no longer have. The problem is still in theory solvable. But it has crossed from a responsive state into a punitive one. Delay changed the nature of the problem.

The same thing happens with health in a more intimate way.

A person starts feeling off. Not movie sick. Not dramatic. Just wrong. Sleep loses depth. Recovery takes longer. Energy feels strangely expensive. The body no longer bounces back from ordinary stress the way it used to. At first this does not feel urgent enough to justify disruption. There is work. There are children. There is rent. There are chores. There is the ordinary social pressure to stay functional and not be dramatic. So the person waits. They hope a weekend will fix it. They hope the feeling will fade on its own. They keep moving.

But the body is moving too.

That is the part people forget. While you delay, the process continues. Blood pressure does not pause because you are busy. Inflammation does not suspend itself out of politeness. Poor sleep does not stop shaping mood, appetite, judgment, and hormones just because the calendar is full. By the time the person finally acts, the same action may still help, but it no longer lands in the same body. What was once a nudge has become a repair. What was once one issue has become several issues reinforcing one another.

Delay does not just postpone the solution. It changes the terrain on which the solution must now work.

That is also true in families.

A child starts struggling in school. At first it may look small. A few missing assignments. A slight withdrawal. A change in mood. Some irritability. Maybe a teacher mentions focus. Maybe the child says everything is fine. Maybe the parent is exhausted and the signs do not yet look big enough to justify alarm. That is understandable. Everyday folks are busy. They live inside pressure. They do not always have the luxury of treating every weak signal like a siren.

But delay changes the chemistry of the problem. A child who starts by needing help can later become a child carrying shame. Shame changes behavior. It creates avoidance. Avoidance creates more missed work. More missed work creates more fear. More fear creates more dishonesty, more hiding, more resistance. Later, the parent is not dealing only with homework. They are dealing with a whole emotional weather system built around the homework. Again, the same category of action no longer works the same way. Help offered early feels like support. Help offered late may be received as pressure, judgment, or proof of failure.

Or think about home maintenance, one of the great teachers of delayed consequences. A small leak under a sink. A dark spot near the ceiling. A tiny crack near a window. The house is still standing. Life goes on. It seems rational to wait until there is more time, more money, more certainty. Then the leak keeps leaking. Wood swells. Paint bubbles. Mold takes hold in places you cannot see. One small problem becomes water damage, ruined drywall, hidden rot, a smell in the walls, a larger bill, and a new kind of stress that now involves not just fixing one pipe, but repairing what time and moisture did while you were waiting.

This is the pattern.

A responsive problem becomes an entrenched problem because time is not empty.

Time is active. Time compounds. Time adds layers. Time recruits new processes into the old process. That is why delay is so dangerous. It does not simply keep things where they are. It allows things to become more organized against you.

The same is true in social life. A hard conversation postponed is not a static conversation. Silence becomes part of the relationship. The other person starts telling themselves a story about your silence. You tell yourself a story about theirs. What was once one painful truth now has a second problem wrapped around it, then a third. Pride enters. Suspicion enters. Habit enters. By the time the conversation finally happens, it has to fight through weeks or months of accumulated interpretation. One delay has created many obstacles.

This is why people so often say, I knew something was wrong, but I did not realize how fast it was becoming harder.

That sentence deserves respect. It points to something precise. Many people are early in a psychological sense and late in an operational sense. They feel the problem before others do. They notice the change in tone, rhythm, pressure, appetite, spending, mood, trust, or energy before the formal world catches up. But feeling early is not the same as having a clean route to act effectively. The person may sense the future while still waiting on permission, money, courage, time, evidence, or support. During that wait, the future does not stand still.

So we need two kinds of early.

Psychologically early means you sensed it before it became official.

Operationally early means you acted while the system was still responsive.

Those are not the same thing.

A person can be psychologically early and operationally late. That may be one of the most common painful facts of adult life. You felt the drift, but not early enough to preserve the easy move. You noticed the body changing, but not early enough to keep the body in its most responsive state. You saw the debt growing, but not early enough to stop interest from building its own machinery.

You sensed the emotional distance, but not early enough to keep the silence from becoming structure.

Once that distinction becomes clear, regret starts looking different. Regret usually tells a brutal fairy tale. It says the signs were there, therefore action should have been simple. But early signs rarely feel simple in real time. They feel uncertain, awkward, socially expensive, easy to dismiss, easy to hope away. Later, once the pattern has become obvious, the mind rewrites the past as if the truth had always worn the same bright uniform. It did not. Earlier, it was softer, weaker, more arguable, and much easier to postpone.

That is why delay is not laziness in a simple sense. It is often a negotiation with ambiguity. It is people trying to avoid paying a cost before they feel fully certain the cost is justified. That is a deeply human thing to do. It is also exactly how intervention windows get lost. Not through dramatic stupidity, but through ordinary postponement inside a world where processes keep moving while people hesitate.

This is what delay really destroys.

It destroys the cheap version of action.

It destroys the clean version of action.

It destroys the moment when a small move could still have made a large difference.

Later, action still exists, but it is no longer the same action in effect. It arrives later, into a thicker problem, at a higher price, with weaker force. That is the practical meaning of losing ground. Not that a person is paralyzed, but that the road between what they do now and what happens next has grown longer and rougher.

That is why delay belongs near the center of this idea. A future can remain visible while the ability to shape it decays. The warning may still be real. The action channel may already be weakening. By the time a problem becomes clear enough to feel undeniable, the world answering that problem may already be less responsive than it was when the first weak signs appeared.

And that is where the deeper question begins.

If delay can do this much damage all by itself, what happens when delay meets noise, friction, bad feedback, and too many moving parts at once.

Delay is only part of the story.

A person can act on time and still discover that their action lands weakly. They can make the call, send the message, book the appointment, cut the spending, tell the truth, change the plan, and still watch the result come back smaller, slower, messier, or stranger than expected. This is where many people become confused. They tell themselves that because they acted, the future should have responded cleanly. When it does not, they assume the action did not matter, or that they themselves must have failed in some simple personal way.

But real life is rarely that direct.

Action does not move through empty space. It moves through resistance.

It moves through noise, friction, delay, misunderstanding, pride, competing incentives, bad timing, social pressure, exhaustion, bureaucracy, habit, and environments that were never designed to cooperate with your best intentions. By the time your present move reaches the future, part of its force may already be gone.

This is one of the biggest reasons people overestimate control when things are easy, and underestimate how quickly control can weaken when life gets complicated. They picture action like a straight beam of light. I decide, I act, the world changes. But in many ordinary situations action behaves more like a message trying to travel through a crowded storm. It gets bent. It gets scattered. It gets weakened. It still moves, but not with the clean power it had at the beginning.

Think about a simple conversation inside a family.

One person decides to finally say what needs to be said. Maybe it is about money. Maybe it is about drinking. Maybe it is about how tired everyone has become. Maybe it is about the fact that one person is carrying too much and has started cracking under the weight. The words themselves may be honest. The timing may look good on paper. The intention may be completely serious. But those words do not arrive in a vacuum.

They arrive through old resentment. Through years of roles and expectations. Through fatigue, shame, defensiveness, fear, and memory. One person hears criticism where another intended concern. One person hears control where another intended love. One person is too exhausted to listen. Another is too scared to admit what is true. The action was real. The conversation happened. But the line between the act and the outcome was crowded. Much of the force got absorbed before it could do its work.

This is friction.

Friction is everything that turns a clean move into a messy one. It is the resistance built into real life.

You see it in health all the time. A person decides they will finally sleep more, eat better, walk more, lower stress, and take their body seriously. A neat decision. A smart decision. A serious decision. Then the week begins. Work runs late. The child gets sick. Bills pile up. Sleep gets broken. The easy healthy meal turns into what can be made in eight minutes. The walk disappears because the weather is bad and the day is full and the body is already tired. The person did not lie. They did decide. But their action had to travel through a life full of drag.

That drag matters. It means the same intention that looked powerful on Sunday night may land weakly by Thursday afternoon.

Or take money. A person decides they are done overspending. They cut subscriptions. They stop ordering food. They cancel the small leaking costs that had become habits. This is good. It matters. But then rent rises. Insurance rises. Groceries rise. A repair appears. A friend needs help. A child needs shoes. The action is still real, but its visible effect gets swallowed by larger pressures. From the inside, this can feel maddening. I made the right move, so why does nothing feel different. The answer is not that the move was fake. The answer is that the move entered a system full of competing forces.

That is dilution.

Dilution is what happens when your action is only one ingredient inside a much larger process.

In a simple situation, one good move can dominate the outcome. Catch the falling glass and the glass does not break. Lock the door and the door stays locked. But in larger situations your action joins many other actions, conditions, and pressures. It may still help, but not in a way that is easy to isolate. A family budget is not just your spending choices. It is wages, prices, debt, timing, emergencies, and the behavior of everyone tied to the household. A relationship is not just one honest conversation. It is history, habits, insecurity, emotional rhythm, outside pressure, and all the interpretations that have piled up over time. A body is not just one good habit. It is a whole ecology of stress, sleep, food, movement, hormones, age, illness, environment, and routine.

This is why people often mistake complicated systems for unfair ones. The real problem is not always unfairness. Often it is just that no single act is large enough to dominate the field anymore.

Another reason action weakens is bad feedback.

Good feedback is one of the great hidden gifts in life. If you act and the world answers clearly, you can learn fast. You know whether the move worked. You know whether to repeat it, adjust it, or abandon it. This is how a child learns to walk, how a cook learns heat, how a musician learns timing, how a gardener learns water and light. The world answers in a way the body can read.

But many adult problems do not answer like that.

You improve your habits for a week and feel no dramatic change, so motivation falls. You try to repair a relationship and the other person seems warmer for two days, then colder again. You cut spending, but the account still looks ugly because the earlier damage is still running through the system. You apply for jobs and hear nothing back. You speak up at work and have no idea whether anyone listened or whether you just put a target on yourself. In situations like these, feedback becomes muddy. You cannot tell what the world is saying back to you.

Once feedback gets muddy, control gets clumsy.

It becomes harder to know whether you are steering or just flailing.

That uncertainty changes behavior. People give up too early. Or they keep pushing the wrong thing too long. Or they chase emotional relief instead of useful response. Or they become superstitious, reading meaning into every tiny shift because the larger pattern refuses to answer cleanly. Weak feedback does not just slow learning. It deforms it.

This is why some of the hardest parts of adult life are not the moments when action is impossible. They are the moments when action is still possible, but the signal coming back is too weak, too delayed, or too contradictory to guide the next move well.

Then there is environmental reinforcement, one of the most underappreciated forces in ordinary life.

A person may truly want to change, but their surroundings keep teaching the old lesson. They want to drink less, but their whole social life runs through bars. They want to spend less, but every app, ad, and habit loop around them is built to trigger impulse. They want to sleep more, but their job rewards constant availability. They want to lose weight, but their stress pushes appetite in the wrong direction and their daily schedule is built around convenience, not care. In every case the person is not merely fighting themselves. They are fighting an environment that keeps feeding the same pattern.

That means action has to work uphill.

And uphill action is expensive.

It takes more effort, more repetition, more clarity, and more recovery to make the same move stick. A choice that would feel easy in a good environment can feel heroic in a bad one. This is another reason people misjudge themselves. They compare their results to an imaginary clean world instead of the world they are actually operating inside.

Noise matters too.

Noise is all the random clutter and variation that makes it harder to tell what is really happening. In a quiet environment, a small signal stands out. In a noisy one, even a good move can disappear. Imagine trying to hear one person speak in a silent room, then trying to hear the same person in a crowded airport. The voice may be equally calm, equally intelligent, equally important. But the setting changes what reaches you.

Life does this constantly.

A person may get a real early signal from the body, but dismiss it because ordinary stress is already making everything feel strange. A couple may try to improve things, but work pressure, child care, money strain, and sleep deprivation generate so much background disturbance that progress becomes hard to detect. A worker may correctly sense that a company is weakening, but daily noise, meetings, short term wins, and forced optimism make the larger signal hard to separate from routine fluctuation.

This is what makes control feel slippery. The signal is there, but the room is too loud.

Once all these forces gather, friction, dilution, weak feedback, environmental drag, and noise, the meaning of action changes. A person can still act, but the same act no longer carries the same force.

A good decision may still be good, yet fail to produce the clean visible shift the person hoped for. This is not because choice is fake. It is because transmission is hard.

That word matters here.

Transmission.

The issue is not only whether you choose. The issue is what happens to the choice after it leaves you. Does it survive the trip. Does it remain coherent. Does it reach the future while the future is still able to answer. Or does it arrive frayed, bent, and weakened by everything it had to pass through on the way.

Once you see life in these terms, many old frustrations become easier to read. You stop asking only whether you acted. You start asking whether your action still had a good road in front of it. You stop assuming that all honest effort should produce neat visible reward. You begin noticing how much of adult life is lived in systems where action is real, but transmission is poor.

And that leads to the next step.

Because once delay and friction have weakened action, and once noise and bad feedback have made the road harder to read, one more force begins to matter more than anything else.

Some problems do not just resist change.

They gather momentum.

There is another reason small problems become hard problems, and hard problems become punishing problems.

They do not simply sit there waiting for you.

They gather momentum.

This is one of the least appreciated facts of ordinary life. People often imagine that if they leave a problem alone, the problem stays in place. Maybe it grows a little. Maybe it becomes more annoying. But in the imagination it is still basically the same thing, just larger. That picture is too gentle. Many problems do not merely become larger versions of themselves. They change character. They recruit new forces. They build history. They start reinforcing their own direction.

That is what momentum means in a human life.

Momentum means a problem begins helping itself continue.

A credit card balance is a simple example. At first the person owes money. That is already unpleasant, but the problem is still mostly arithmetic. Spend less, earn more, pay it down, move on. Then interest starts doing its slow work. One missed payment becomes a fee. The fee becomes a heavier statement. The heavier statement creates stress. Stress weakens discipline. Weak discipline invites another bad month. Another bad month increases the balance again. Now the person is no longer facing one simple number. They are facing a loop. The debt is not just sitting there. It is shaping the behavior of the person who is trying to escape it.

That is how a static problem becomes a moving one.

The same pattern appears in the body.

Poor sleep is rarely just poor sleep for long. At first a person is tired. Then the tiredness changes appetite. Appetite changes energy. Energy changes movement. Movement changes mood. Mood changes food choices. Food choices change inflammation. Inflammation changes sleep again. What began as one weak disruption becomes a network. The person wakes each day and feels as if the whole body has quietly taken the side of the problem. That feeling is not imaginary. Systems do build momentum. The body starts adapting around the disturbance. What was once a soft warning becomes a hardened pattern.

This is why late action feels so discouraging. A person does not merely face the original issue. They face the extra machinery the issue built while it was allowed to continue.

A marriage can work the same way. At first there is disappointment. One person feels unheard. Another feels criticized. Neither wants a war, so both become careful. But careful in the wrong way. They stop saying the risky truthful thing. They start using tone instead of language. Small injuries remain unresolved. Unresolved injuries become stored memory. Stored memory changes interpretation. Once interpretation changes, every new event begins passing through an older story. A harmless delay now feels like rejection. A simple request now feels like accusation. A tired mood now feels like contempt. At that point the couple is no longer arguing about one forgotten dinner, one rude sentence, one bad week. They are living inside accumulated meaning. The relationship has built momentum in the wrong direction.

This is why some people say things like, we have not been fighting about the present for a long time. We have been fighting about the archive.

That sentence, in everyday language, is exactly what momentum feels like.

The past starts pushing on the present.

Habits do this too, often with embarrassing efficiency. A person skips exercise for a week. Then the body feels heavier. Feeling heavier makes movement less attractive. Less movement changes sleep and mood. Worse sleep and lower mood make comfort feel urgent. Comfort leads toward overeating, scrolling, drinking, avoidance, or staying up too late. Then shame appears. Shame does not usually improve discipline. Shame often drives more avoidance. Soon the person is not dealing with one lazy week. They are dealing with a whole pattern that now has emotional gravity. The habit has acquired companions. It has mood, routine, bodily memory, and self story on its side.

This is the real cruelty of momentum. It colonizes the rest of life.

One weak point invites another.

One delay makes the next delay easier.

One compromise changes the standard for the next compromise.

One tolerated discomfort becomes the environment in which future decisions are made.

That is why early correction matters so much. Not because early correction is morally pure, but because it interrupts the problem before the problem can build allies.

A small water leak, if fixed early, is a plumbing issue. Left alone, it becomes damaged wood, stained ceilings, mold, strange smells, hidden rot, ruined paint, and eventually a feeling that the whole house is slightly sick. The leak brought friends. It became more than itself. A person who addresses it later may still solve it, but now they are no longer solving one problem. They are solving the ecosystem the problem created.

This is the shape of hardening.

Hardening means the future stops being soft enough to answer small corrections.

At first, a small push changes direction. Later, the same push barely registers.

Think about weight gain. Five pounds is one thing. Twenty five pounds plus lower energy, plus worse sleep, plus back pain, plus lower confidence, plus habits built around convenience, plus a day organized around sitting, is something else entirely. The later situation is not just the earlier situation with more numbers. It is a hardened arrangement. More of life has adapted around it. The body has adapted. The schedule has adapted. The emotions have adapted. The identity may even have adapted. A person can still change, of course. But they are no longer pushing on the same kind of door.

Or consider a child who starts avoiding schoolwork. Early on, the issue may just be confusion or distraction. A tutor, a calmer evening, a little support, a good teacher, and the problem may loosen quickly. Later, after months of missed work, embarrassment, parent conflict, fear of being exposed, and the quiet panic of falling behind, the issue is not just learning. It is learning plus shame. Shame hardens problems. Shame adds secrecy. Secrecy slows help. Slower help leads to more failure. By then the child is no longer resisting homework alone. The child is resisting the whole emotional pain attached to homework.

This is why the later stage of many problems feels strangely overbuilt. The person says, I do not understand why something so small now feels so huge. The answer is that it is no longer small, even if it started small. Time has given it structure. Repetition has given it muscle. Avoidance has given it room. The problem learned the house.

There is also a public version of this. A company can drift the same way. At first morale dips a little. Then people stop trusting leadership. Once trust weakens, good workers leave. Once good workers leave, the people who remain carry more pressure. More pressure creates more mistakes and more resentment. Resentment reduces care. Reduced care lowers quality. Lower quality weakens the business further. Now the organization is not suffering one bad quarter. It is suffering a chain reaction. By the time the decline is visible to everybody, much of the damage is no longer sitting in spreadsheets. It is sitting inside culture, habit, and expectation. That too is momentum.

One of the hardest truths here is that clarity often rises with hardening. People sometimes think they finally understand the problem because they have become wiser. Maybe they have. But often the problem also became easier to read because it became more committed to its path. A marriage in deep coldness is easier to diagnose than a marriage in early drift. Serious debt is easier to name than mild overspending. Full burnout is easier to recognize than subtle depletion. A failing company is easier to describe once the departures and cuts are impossible to hide. In each case the signal got stronger, but the signal got stronger partly because the situation got stiffer.

That is why late clarity has such a bitter taste.

The problem is obvious now, but obvious now is not the same as easy now.

This is the sentence people need.

Obvious now is not the same as easy now.

It explains why hindsight feels so unfair. Looking back, a person sees a neat line where in real time there was only drift, ambiguity, hope, and weak signal. Looking back, they think the answer should have been clear. But earlier the answer was mixed with uncertainty, and later the answer was mixed with momentum. The mind confuses stronger evidence with stronger opportunity. Often the opposite is true. Stronger evidence can mean weaker opportunity, because the process has had more time to organize itself.

This changes how we should think about timing. Timing is not only about catching the beginning. It is about acting before the system becomes too committed to itself. Before the debt recruits interest and shame. Before the body recruits more loops. Before the relationship recruits interpretation and distance. Before the habit recruits identity. Before the institution recruits cynicism. Before the house recruits mold.

The reason this matters so much is simple. Momentum is expensive.

It raises cost, but it also lowers responsiveness.

That second part is easy to miss. People focus on expense because expense is visible. Bigger bills, bigger fights, bigger stress, bigger damage. But responsiveness may matter even more. A system with low responsiveness still absorbs action, but it no longer yields much from small interventions. You can push, and push honestly, and still feel as if the world is barely moving.

This is where so many people start saying I waited too long.

What they usually mean is not just that they delayed. What they mean is that they allowed the problem to acquire momentum. They let it cross from a state where the world still answered lightly into a state where the world answers heavily. They are no longer negotiating with a disturbance. They are negotiating with a trajectory.

And once a trajectory has formed, the most important question in life becomes impossible to ignore.

How much practical reach is left from here.

Because that is the point where the problem stops being only about warning signs and starts becoming about the boundary itself. The place where your actions, even good ones, no longer land with the dependable force they once might have had.

There comes a point in many problems when the old kind of effort no longer works the way it used to.

That point is not always dramatic. There is not always a loud moment when the world announces, from now on, your choices will matter less. More often it feels ordinary. A person tries the same type of move that might have helped months earlier, and the result comes back smaller. Slower. Weaker. More expensive. Less reliable. The future is still there. The person is still there. Action still exists. But the connection between what they do now and what happens next has changed.

That is the point we need to name.

The Agency Horizon is the boundary where a person still acts, still cares, still tries, but the action no longer reaches the future with enough dependable force to count on.

Not because all hope has vanished. Not because reality has become magical. Not because choice is fake. Simply because the road between present effort and later result has become too long, too delayed, too noisy, too crowded, or too stiff to answer cleanly.

This is easier to understand with plain examples.

Think about a person trying to save money. Early on, small changes can work quickly. Canceling wasteful subscriptions matters. Cooking at home matters. Cutting a few habits matters. The numbers answer back. The person feels that their decisions still reach the near future. They do something in the present, and the next month begins to look different.

Now imagine the same person much later, under high interest debt, late fees, rising rent, and a car problem that cannot be postponed anymore. They can still cut spending. That is still wise. But the same action no longer produces the same effect. The field around the person has thickened. Their effort now enters a harsher arrangement. Money saved is swallowed by obligations already in motion. The act remains real, but its reach has weakened. At some point the problem crossed from manageable pressure into a region where ordinary corrective action no longer works reliably enough to restore control. That crossing is a version of the Agency Horizon.

Or take a body under stress. Early on, one week of better sleep, less alcohol, more walking, and a quieter schedule may produce a noticeable difference. Energy rises, mood steadies, appetite softens, focus improves. The body is still responsive. Present effort still reaches the near future strongly. But later, after years of overload, chronic sleep loss, poor food, high blood pressure, weight gain, and low recovery, the same good habits may still help, yet they do not land with the same clean force. A walk helps, but not much. One good night helps, but not enough. A better week helps, but does not restore the old self. The body has crossed into a different region. The person is no longer pushing on the same door. This too is a version of the Agency Horizon.

The same pattern appears in human bonds. Early in a conflict, one honest conversation can still change the emotional weather. A good apology can clear tension. A difficult truth can prevent months of poison. But later, after resentment has settled in, after silence has become routine, after both people have been privately rewriting the relationship in their own minds, the same conversation no longer carries the same power. The words may be good. The intent may be sincere. Still, the effect is weaker, slower, and harder to trust. A bond that was once soft enough to answer has become guarded. The people are no longer speaking into open air. They are speaking through months of accumulated meaning. Again, action remains possible. But dependable influence has fallen.

This is the important distinction.

The Agency Horizon is not the point where action becomes impossible in principle.

It is the point where action stops being something you can rely on to work in the ordinary way.

That difference matters because people often torture themselves with impossible standards. They tell themselves that as long as some action was theoretically available, they should have been able to fix the problem. But ordinary life is not judged by theory. It is judged by practical reach. A person does not need infinite choice. They need enough clean contact with what happens next. Once that contact weakens past a certain point, the emotional meaning of effort changes. Action becomes more like damage control than guidance. More like recovery than steering. More like surviving impact than preventing impact.

This is why the horizon idea is so useful. It stops us from asking only, could someone still do something. That question is too crude. Almost always, someone can still do something. They can still send the message, make the call, cut the cost, tell the truth, change the habit, seek help, apologize, leave, stay, or begin again. The sharper question is whether those actions still have a good chance of changing the path in a dependable way from where the person now stands.

That is the real test.

Can this present move still reach the future with enough strength, enough speed, and enough clarity to matter reliably.

If the answer is yes, agency remains strong.

If the answer is no, the horizon is near, or has already been crossed.

You can feel this in ordinary repairs. A small roof leak caught quickly is a routine problem. The same leak left alone becomes stained ceilings, warped wood, mold, ruined insulation, and a house that now smells wrong. Calling the roofer later is still better than never calling. But the act belongs to a different category now. It no longer prevents damage. It enters after damage has already spread. The present action still matters, but it has lost its old commanding position. It now lands in a field that has become less responsive.

You can feel it in school as well. A child who needs support early may respond quickly to structure, attention, and calm help. The same child, after months of falling behind, shame, and avoidance, may still be helped, but the path is harder. The issue is no longer only reading, or focus, or homework. The issue now includes fear, identity, self protection, and emotional fatigue. Help still matters. But it arrives into a more hardened life. The old easy leverage is gone.

This is why the horizon is often missed in real time. It is not painted on the wall. People usually realize it only after the fact, when they look back and notice that the same kind of move had once carried much greater force. Earlier, the issue answered quickly. Later, it answered slowly or badly. Earlier, one good step changed the whole week. Later, ten steps barely changed the day. Earlier, the future was still close. Later, it had drifted farther away.

And this is where many of the worst emotional mistakes begin. A person crosses the horizon without naming it, then keeps applying the same small moves, hoping for the same old response. They treat a hardened problem as if it were still a fresh one. They offer one apology to a bond that now needs a season of repair. They make one budget change against debt that now behaves like a machine. They give one weekend of rest to a body that now needs an entirely different life. They are not foolish for trying. They are acting with an outdated map.

The Agency Horizon is what updates the map.

It tells us that timing changes the meaning of action. It tells us that effort is not judged only by virtue, but by contact. It tells us that a visible future may still be out there, clear enough to read, while the line from present effort to later change has already grown weak.

This is also why waiting for certainty can be so dangerous. Certainty often arrives late. By the time the problem becomes undeniable to everyone, the horizon may already be behind you. The evidence is stronger, but the leverage is weaker. The pattern is clearer, but the future is harder to move. That is one of the bitter little laws of ordinary life. Better proof can arrive together with worse reach.

Once a reader feels that, the next step becomes unavoidable.

Sometimes the cruelest part is not only that leverage fades.

Sometimes the future remains readable even after leverage has already begun to die.

That is the next boundary we have to face.

Once people understand this boundary, a great many memories stop looking random.

They begin to sort themselves into a pattern.

A person remembers the tooth that only hurt a little at first, then later owned the whole side of the face. A person remembers the strange engine noise that seemed harmless until the car died in the wrong place at the wrong time. A person remembers the credit card bill that felt annoying, then became a second job made of panic and minimum payments. A person remembers the marriage that did not collapse in one night, but thinned and cooled and stiffened until every conversation felt late. A person remembers the body asking softly for help, then speaking more loudly, then finally refusing to be ignored.

What changes in all these memories is not only the size of the problem. What changes is the meaning of action.

Earlier, action could still guide the future.

Later, action had to chase the future.

That difference is enormous.

When a problem is still before the horizon, even a modest move can matter a great deal. A filling instead of a root canal. A mechanic visit instead of a tow truck. A budget correction instead of long term debt pressure. One honest conversation instead of a year of coldness. A season of rest and discipline instead of a body dragged into long repair. The future is still close enough to answer. You touch it, and it moves.

After the horizon, the same kind of effort may still be wise, but it no longer has the same authority. The filling becomes a root canal. The mechanic visit becomes a major repair bill. The budget correction becomes survival. The honest conversation becomes a painful excavation through layers of history. Rest becomes treatment. Treatment becomes management. Management becomes the best available version of a situation that earlier might have responded to far less.

This is why people so often describe their lives in terms of being overtaken.

They do not always mean that the problem was invisible. Often they mean something subtler. They mean the problem became more organized than they were. It built more momentum than they did. It gathered more structure, more cost, more reinforcement, more reality, than their current action could handle cleanly.

That is the practical meaning of crossing the horizon.

The world starts answering late, weakly, and at high cost.

This should immediately change the kind of questions a person asks. A great many people waste years asking the wrong one. They ask, can I still do something. That question sounds reasonable, but it is too blunt. In almost every hard situation, some action still exists. You can still call. You can still leave. You can still cut spending. You can still apologize. You can still go to the doctor. You can still ask for help. You can still stop lying to yourself. The mere existence of action is not the real issue.

The real issue is this.

What kind of action still has dependable contact with the future from here.

That is a much better question.

It is a question that respects timing. It respects cost. It respects momentum. It respects the difference between a soft system and a hardened one. It respects the reality that a future can remain visible while becoming harder to touch.

Once that question becomes central, life looks different.

A person with rising debt stops asking whether a heroic turnaround is possible in fantasy, and starts asking what can still be reliably changed this month, with this income, under this pressure, before things get worse. A couple in trouble stops pretending that one grand talk will erase a year of distance, and starts asking what kind of repair is still real, what trust can still be rebuilt, what damage has become permanent, and what honesty is now overdue. A tired overworked person stops asking whether one good weekend will restore the old self, and starts asking what changes in schedule, responsibility, food, sleep, and stress are actually large enough to matter now. A parent stops asking only whether a child can still catch up someday, and starts asking what support still lands while shame has not yet hardened further.

These are painful questions. They are also adult questions.

They move a person away from fantasy and toward contact. Away from pride and toward timing. Away from the false comfort of saying, I still have options, and toward the more exact truth, these are the kinds of options that still work from here, and these are the ones that now belong mostly to imagination.

There is mercy in this, even though it does not feel soft.

The mercy is not in pretending every lost window can be reopened on demand. The mercy is in stopping the endless self torture that comes from using the wrong map. A person who has crossed the horizon and still judges themselves by the standards of an earlier softer situation will only deepen their suffering. They will keep demanding from late effort the kind of clean response that only early effort could have received. They will keep calling themselves weak because reality no longer answers in a simple way. They will keep confusing changed terrain with personal failure.

A better map does not erase pain, but it changes the shape of blame.

It allows a person to say, the future was not equally reachable at every stage.

That sentence may sound small, but it carries a lot. It tells the truth about timing. It tells the truth about why a weak signal ignored becomes a stronger signal in a less forgiving world. It tells the truth about why hindsight is cruel, because hindsight sees the hardened pattern and forgets the softer uncertainty that came before it. It tells the truth about why so many people live with the sentence I knew, and still feel haunted. Often they did know something. But knowing is not the same as having strong enough contact with the future to steer it well.

This is where the horizon idea becomes more than a concept. It becomes a way of reading ordinary life.

You stop thinking only in terms of events.

You start thinking in terms of windows.

There was a window where the body still answered lightly. A window where the bond still answered honestly. A window where the debt still answered arithmetic. A window where the child still answered support before shame became part of the lesson. A window where the house still answered repair before rot recruited mold and damage spread through hidden spaces.

Most of adult life is not about avoiding every problem forever. That is impossible. Most of adult life is about learning to notice which problems are still in their soft phase, and respecting them before they recruit time to their side. That is one of the deepest practical lessons in this whole line of thought. The earlier stage of a problem often feels less dramatic, but it contains more freedom. The later stage often feels clearer, but it contains less.

This is why waiting for certainty is so dangerous. Certainty often arrives as the horizon closes. The evidence becomes stronger, but the leverage becomes weaker. The truth becomes easier to name, but harder to move. By the time everyone agrees that the problem is real, the problem may already have built enough momentum to resist ordinary correction.

Once a person really feels that, another realization appears.

Sometimes the cruelest thing is not merely that action weakens.

Sometimes, even after action has begun losing its force, the future keeps giving off signals. The pattern keeps becoming clearer. The direction keeps becoming more legible. The world keeps telling you what is likely, even after your strongest chance to reshape what is likely has already begun to fade. That is a darker and more precise condition.

It means the horizon does not end visibility.

It only ends dependable leverage.

And that leads to the next step. Because once you can see that action can fall below useful strength while prediction remains alive, you are standing in front of one of the most painful structures in ordinary life. You can still read the future.

But your hand no longer reaches it the way it once did.

That is where this path now goes. Into the region where knowledge remains, warning remains, pattern remains, and yet practical leverage has already started to die.

Relationships are one of the most painful places where this split appears, because love does not usually fail like a light switch. It fails like weather. Pressure changes first. Temperature changes first. The room feels different before anyone agrees on why. A tone enters the conversation that was not there before. Jokes land a little flatter. Curiosity drops. Tenderness becomes less spontaneous and more procedural. Replies come, but they do not carry the same warmth. The body still sits at the table, the messages still arrive, the daily routine still exists, but something inside the bond has already begun to loosen.

This is why so many people say they knew long before the ending became official. They say they felt it in the pauses, in the way affection began to sound effortful, in the way certain subjects stopped opening and started closing. They say nothing big had happened yet, but the emotional climate had changed. That is a very precise description of a relationship becoming legible before it remains easy to save. The pattern is visible before the handle feels solid. The future enters the room before either person is ready to speak its name.

That early stage is one of the strangest emotional zones in human life. It is not ignorance, but it is not proof. It is not peace, but it is not yet collapse. A person starts living with a quiet double awareness. On the surface, life continues. Plans are still discussed. Bills are still paid. Ordinary routines are still performed. But underneath that surface, perception has already updated. Something has shifted in the way contact works. The relationship may still exist in form, while already weakening in signal.

This matters because relationships are not moved by abstract truth alone. They are moved by timing, honesty, mutual reach, and the condition of the loop between two people. Once that loop degrades, action becomes much harder. A conversation that might have repaired something six months earlier may become mostly diagnostic six months later. An apology that once could have reopened trust may land as a summary of damage already done. A request for closeness that would once have felt natural may later feel invasive, desperate, or misaligned with the emotional reality that has already taken hold. This is exactly what Future Access Without Presence looks like in intimate life. You can often see the direction before you can still easily change it.

The painful thing is that relationships are full of ambiguous signals, and ambiguity weakens action. If a person had a clean laboratory readout of a bond, a clear meter on trust, warmth, resentment, attention, and hope, then timely action would be much easier. But love does not present itself that way. It arrives through mood, rhythm, tone, silence, softness, and omission. It speaks through what no longer happens as much as through what does. This makes the early stage of trouble very difficult to act on. A person worries about overreacting. They worry about sounding needy, dramatic, paranoid, or insecure. So they keep collecting evidence, not realizing that the very act of waiting may be letting the emotional field harden underneath them.

That hardening is one of the central facts of relationship life. Feelings do not simply rise and fall in isolation. They form habits. Distance becomes patterned. Resentment becomes practiced. Withdrawal becomes easier. Once a couple spends enough time relating through politeness, exhaustion, or low grade disappointment, that style of connection can become the new normal. At that point the problem is no longer only the original wound. The problem is that the bond has learned a new equilibrium. That is why late interventions often feel so heartbreaking. Both people may finally be ready to admit something is wrong, but they are now speaking inside a structure that has already adapted to wrongness.

Think about how often people say the breakup seemed sudden from the outside. Friends are shocked. Family members say they had no idea. Social media looked normal. The photos were still there. The routines still appeared intact. But from the inside, the ending may have been visible for a very long time. The future was leaking long before the event. One or both people were living in that long middle zone where affection was thinning, repair was slowing, and hope was becoming more ceremonial than alive. The official end was not the beginning of the process. It was the moment the process became undeniable.

That gap between inner timing and outer timing creates a lot of suffering. It leaves people feeling strangely alone even before they are technically alone. They are already reacting to a future that other people cannot yet see. They are already grieving something that has not publicly ended. They are already trying to decide whether what they feel is real enough to justify a difficult conversation. That is one reason relationship strain is so exhausting. It forces a person to make decisions under conditions of partial visibility. They have signal, but not certainty. They have pattern, but not permission. They have awareness, but not always a clean route to leverage.

The digital age has made this even more complicated, not simpler. People now have endless surfaces of contact, but surfaces are not the same as closeness. A couple can exchange messages all day and still live inside emotional distance. They can maintain the appearance of continuity while the substance of connection erodes. Read receipts, typing bubbles, delayed replies, short phrases, reactive emojis, and sparse affection all become tiny carriers of meaning. This creates a kind of hyper legibility at the level of signal fragments, while often doing very little to restore real contact. In some cases it makes the problem worse. People become expert readers of symptoms while becoming worse at entering the direct conversations that might still matter.

That is another relationship version of modern Future Access Without Presence. You can know a lot about the changing pattern of a bond without being any closer to repairing it. You can detect warmth loss in high resolution and still be unable to convert detection into trust. You can monitor the decline without interrupting it. In fact, endless monitoring often becomes its own trap. A person studies every reply, every pause, every tonal shift, every absence of enthusiasm, and calls this understanding the relationship. But understanding the decay is not the same as changing the conditions that produce it.

This is why the most important moments in intimate life often happen earlier than people want them to happen. The useful conversation is usually not the conversation that takes place after all doubt has been removed. By then both people may simply be naming the weather. The useful conversation is the one that takes place when the signals are still weak enough to be deniable and strong enough to be real. That is the difficult zone because it demands action before certainty. It asks a person to risk looking foolish in order to remain in contact with a future that has not fully closed yet.

Many people fail there, not because they do not care, but because the emotional cost of early honesty is high. To speak too soon risks embarrassment. It risks hearing that nothing is wrong. It risks making the relationship feel more fragile by naming fragility. It risks exposing need. It risks forcing a definition before either person feels ready. So people wait. They try to be patient. They try to be reasonable. They tell themselves that the mood will pass, that stress is temporary, that work has been intense, that everyone goes through phases, that love naturally changes shape. Sometimes those explanations are true. But sometimes they function as delay mechanisms, stories that allow the bond to keep drifting while the part that is still reachable gets smaller.

This is why relationships teach the FAWP lesson with such cruel precision. Love is not only about feeling. It is also about timing under ambiguity. You do not need a final verdict to know that something essential has changed. You need enough honesty to recognize when warmth is no longer self renewing, when trust is no longer repairing itself, when avoidance has become a climate instead of a phase. Those are the moments when leverage still exists, but it rarely feels clean or heroic. It feels awkward, exposed, and premature. That is exactly why it matters.

And even when a relationship cannot be saved, this framework still helps explain the pain. People often torture themselves by believing that if they saw the signs, they must have been able to stop the ending. That is not always true. Sometimes the bond was already moving beyond their reach by the time it became emotionally visible. Sometimes the conversation could only clarify, not reverse. Sometimes the most that remained possible was a cleaner exit, a less cruel truth, a less destructive delay. That may sound small, but it is not small at all. In intimate life, even limited leverage matters. The difference between honest timing and late denial can shape the emotional cost of an ending for years.

That is what makes relationships such a powerful case. They show that the future can arrive first as atmosphere. They show that clarity can grow while influence shrinks. They show that people often begin to live inside the truth of an ending long before they are ready to announce it. And they show, with unusual tenderness and brutality, how much of human suffering comes from the gap between feeling the direction and still having enough shared contact to change it.

Chapter 4

There is a particular kind of pain that does not come from being surprised. It comes from being right too early, and still being unable to do much with it.

Most people assume the worst moments in life are the ones where disaster arrives out of nowhere. The diagnosis you never saw coming. The breakup that hits like a car crash. The market collapse that seems to appear in a single morning. The layoff email that drops into your inbox like a bomb from a clear sky. Those moments are dramatic, and they are real, but they are not the only kind of pain that matters. In many cases, the deeper wound is stranger than that. You do see it coming. You feel the angle of it. You notice the drift. You sense the pattern before the official event arrives. The future does not stay hidden from you. It stands there in outline, visible enough to make your stomach tighten. The problem is that seeing the outline does not mean your hands still connect to the machinery. That is one of the cruelest truths in modern life. The future can remain readable after your power to change it has already started to weaken.

This is the territory we need to name carefully, because people live inside it all the time and often blame themselves for it. They say they should have acted harder, spoken sooner, moved faster, believed themselves more, trusted their gut, pulled the trigger, made the trade, left the relationship, gone to the doctor, quit the job, called the bluff, changed the pattern. Sometimes that is true. Sometimes earlier action really would have helped. But not always. Sometimes the deeper problem is not lack of character. Sometimes the deeper problem is that prediction and power are not the same thing, and they do not fade at the same speed.

A pattern can stay visible even while your influence over it collapses.

That sentence matters because it pulls apart two ideas that people constantly fuse together. The first is legibility. Can you still read where things are heading. Can you still detect the direction of motion. Can you still feel the curve of the process unfolding around you. The second is leverage. Can anything you do from here still alter the outcome in a reliable, meaningful, timely way. Those sound similar when life is simple. In a small, clean situation with fast feedback, they often move together. You see the problem, you act, the world answers back quickly enough for you to correct course. But modern life is full of situations that are not small, not clean, and not fast. The world has more delay, more noise, more friction, more distributed action, more crowded channels, more sluggish response, more half-signals, more false starts. In that kind of world, reading and steering begin to separate.

That separation is the heart of what we are trying to understand.

Think about a relationship that is ending. In movies, the breakup often arrives as a speech. One person says the line, the scene lands, the music changes, and both people know that the story has crossed a border. Real life is usually much less theatrical. The shift is often visible far earlier than the formal ending. The rhythm changes first. Replies become thinner. Laughter starts arriving late. Tiny gestures disappear. The atmosphere cools in ways that are hard to prove but easy to feel. The center of the bond begins to hollow out long before anyone says the words out loud. A person may recognize this months in advance. They are not guessing randomly. They are reading a real pattern. But reading the pattern does not automatically restore warmth, rebuild trust, erase resentment, or reopen a window that has already narrowed. The future of the relationship can be visible before it is still reachable.

The same structure appears in work. Long before the official notice comes, people often know that something is dying. The language around the project changes. Decisions start drifting elsewhere. Energy drains from the room. Commitments become strangely vague. No one says the thing directly, but the pattern is already there. A person may see the decline clearly, may even describe it accurately to a friend over dinner, and still have almost no meaningful way to reverse it. They are not blind. They are not passive. They are inside a system whose direction has become easier to notice than to alter.

Health can work this way too. Many people know in some quiet, uneasy sense that something is off before any label exists. The body is not a courtroom. It does not wait to present a complete formal case. It leaks hints. Stamina changes. Sleep changes. Pain migrates. Appetite shifts. Mood bends. The person notices. They are not inventing the pattern out of fantasy. But noticing does not guarantee a timely path to intervention. Sometimes the signal is real before medicine can act on it cleanly. Sometimes the condition becomes legible before it becomes tractable. The body can speak before leverage arrives.

And then there is finance, which may be one of the purest everyday theaters of this whole pattern. People talk as if seeing a crash coming and profiting from it were nearly the same thing. They are not. A market can look unstable, stretched, exhausted, absurdly priced, or structurally fragile long before a clean, profitable, survivable trade becomes available. There is a massive difference between reading danger and extracting usable advantage from it. Timing can destroy you. Volatility can shake you out. Crowding can erase the edge. Carry can bleed you dry. A market can stay irrational longer than your position stays alive. You can be right about the direction and still lose in the only currency that matters, which is reality. Finance is full of people who confuse prediction with control, and the market punishes that confusion with unusual elegance.

That is why this subject matters so much. It is not only about knowledge. It is about the geometry of action.

When people say, I knew it was coming, they are often describing something real. The mistake is assuming that this sentence secretly means, and therefore I could have stopped it. Sometimes those two statements line up. Often they do not. One refers to the information channel. The other refers to the intervention channel. The information channel asks whether the present contains usable clues about what comes next. The intervention channel asks whether what you do now still has enough timely contact with the future to bend it. In many real environments, the first channel survives longer than the second.

This produces one of the most maddening emotional states a person can experience. Not ignorance, but clarity without reach. Not confusion, but recognition without traction. Not blindness, but late contact with a process that has already built momentum.

That pain has a distinct texture. It creates regret, but a particular kind of regret. Not just, I should have known. More often, I did know, but I did not know what could still be done from here. Or worse, I sensed it, but I could not tell whether the window was still open. That uncertainty matters. In real life, windows do not usually come with timers floating above them. There is no glowing countdown in the sky announcing that your leverage expires in eleven days, or six meetings, or three more missed signals, or one more quarter, or one more margin call. Human beings must estimate the state of the window from inside the situation itself, with incomplete information, emotional bias, social pressure, noisy feedback, and often very bad timing.

That is why this should not be reduced to a morality play. People love simple stories about courage. Speak up. Act sooner. Trust yourself. Be decisive. Those lessons have their place, and sometimes they are exactly right. But they are not the whole story. There are many situations in which the issue is not raw bravery. The issue is that the system has already drifted into a region where the pattern is still visible but the control surface has gone soft. You can turn the wheel and feel almost nothing answer back.

That image is worth holding onto. Imagine driving a car where the windshield remains perfectly clear, but the steering response becomes delayed, mushy, and uncertain. The road ahead is still visible. In fact, it may become even more visible because danger sharpens your attention. But the clarity of the view does not restore the quality of control. Seeing the bend in the road does not guarantee that your input reaches the tires in time. That is the cruel split. The map still works. The controls do not.

Much of modern life feels like that. We have more dashboards than any generation before us. More metrics, more updates, more commentary, more analysis, more signals, more warnings, more trending lines, more stories about trajectories. In some ways we are flooded with legibility. We can often see patterns sooner than people in the past could. But that does not mean our local power has grown in proportion. In many environments, it has not. The systems are too large, too fast, too mediated, too delayed, too distributed. Visibility rises. Leverage does not keep up.

This is why the coming pages matter. We are not talking about prophecy. We are not talking about magical foresight. We are not talking about certainty, and we are not declaring that nothing can ever be changed. The point is more precise than that, and more useful. There are situations in which the future remains detectable while the action channel degrades below practical strength. There are situations in which knowing survives after steering begins to fail.

Once you see that, a great many ordinary experiences start to reorganize themselves. The breakup that felt obvious but unstoppable. The job that felt doomed before the paperwork. The body that warned before there was a clean path to act. The market move that looked visible but stayed untradeable. The social process that seemed clear but could not be redirected from any one local position. These are not all the same story, but they can share the same structure.

The future can remain legible after leverage is already dying.

That is the paradox we need to follow, because hidden inside it is one of the most important facts about life in a delayed, noisy, high-friction world.

To see why this happens, you have to stop thinking of knowledge and influence as two names for the same thing. They are not twins. They are different channels, and they age differently.

A pattern can remain easy to read for the same reason a scar remains easy to see. Once a process has built enough structure, it starts leaving traces everywhere. It shows up in behavior, in timing, in prices, in tone, in delays, in repeated mistakes, in shrinking options, in little distortions that begin to line up. The world starts giving off clues. It begins to rhyme with itself. A person does not need supernatural powers to notice that. They only need enough contact with the pattern for the pattern to become legible.

Influence is harder. Influence does not just require awareness. It requires contact of the right kind, at the right time, with enough strength, and with feedback that arrives soon enough to matter. That is a much stricter demand. A future can be readable from a distance. To change it, your action has to get inside the machinery while the machinery is still sensitive to being changed.

That difference sounds technical at first, but it is actually very ordinary. Think about a spill on a kitchen floor. Early on, you can stop it with a cloth. A small movement is enough. A few seconds later, it has spread. Now you need more effort. A little later, it has seeped under things, reached edges, found grooves, started soaking into materials. The direction of the spill may be clearer than ever. You can see exactly where it is going. But the effort needed to change the result has grown quickly. Visibility rises as control decays. The pattern becomes more obvious at the same time that the intervention becomes worse.

A lot of life works like that.

This is one of the simplest reasons the leverage gap exists. Structure tends to accumulate clues. Action tends to lose efficiency. The two curves do not have to move together. In fact, they often move apart.

Markets show this very clearly. A bubble rarely looks most normal right before it breaks. Usually the signs pile up first. Prices detach from plain reality. Narratives become too smooth. The same slogans repeat everywhere. People stop asking basic questions because the emotional weather punishes doubt. The structure becomes easy to recognize, especially to anyone watching with a calm eye. But recognizing the instability is not the same as having a clean way to act on it. You can short too early. You can get squeezed. You can bleed while waiting. You can be right in theory and destroyed in practice. The pattern becomes visible before action becomes clean.

That is the deeper lesson. In many systems, prediction survives because the process keeps broadcasting itself. Steering dies because action must survive timing, cost, resistance, uncertainty, and weak coupling.

Coupling is one of the most important ideas here. It simply means the strength of the link between what you do and what happens next. If the link is strong, a small move can still matter. If the link is weak, even a serious effort may vanish into the background. People often imagine action as a dramatic moral event. You decide, you act, the world changes. But real systems are rarely so polite. Many of them are thick. They absorb force badly. They blur signals. They answer late. They answer ambiguously. They answer in ways that make learning harder. You push, but it becomes unclear whether anything moved because of you or in spite of you or not at all.

This is especially painful in human relationships, because people naturally interpret weak coupling as a failure of sincerity. They think, if I really meant it, it would have worked. But that is not always true. A relationship can enter a state where the signs are easy to read and the fixes are hard to deliver. The problem is not that one person suddenly becomes blind. The problem is that trust, once thinned out, stops transmitting repair efficiently. The same words that might once have helped now arrive late. The same gestures carry less force. The bond still produces information, but it carries action poorly. The future of the relationship remains readable even while the channel for changing it weakens.

Health offers another version of the same structure. Early warning signs are often vague, but not meaningless. Fatigue, inflammation, pain, sleep disruption, strange rhythms, altered recovery, these can all be real clues. Over time, the pattern may become easier to see. Symptoms may align. A person may feel, with growing confidence, that something is wrong. But the body is not a system where greater visibility always means easier intervention. Sometimes it is the opposite. The clearer the pattern becomes, the more it reflects the fact that the process has advanced. The evidence grows at the same time that leverage shrinks. That is why so many people feel haunted by their own hindsight. They were not imagining the pattern. They were detecting it. The cruelty is that detection did not guarantee a workable opening.

There is a broad principle here. Prediction often benefits from persistence. If a process has momentum, the present tends to contain clues about the future. Steering depends on sensitivity. If the process has hardened, accelerated, spread out, or locked itself into reinforcing loops, the future may still be visible while becoming harder to bend.

That is the real asymmetry. Legibility feeds on pattern. Influence feeds on responsiveness. A system can have one without the other.

Weather is a simple example. A storm system can be tracked days in advance. Its direction, strength, and probable landfall can become increasingly clear. Forecasting improves as the structure of the storm becomes legible. But no one mistakes that for personal control. You can read the storm, prepare for the storm, and survive the storm, but not steer the storm. The information channel is strong. The intervention channel is practically absent. In everyday life, many situations are not as absolute as weather, but the shape is similar. The future can be progressively easier to read while remaining stubbornly difficult to change.

Work and institutions often produce this feeling in a quieter way. People can sense decay long before there is a formal announcement. A project loses energy. Standards slip. key people withdraw their attention. The language around the future becomes thinner and less confident. The process starts leaking its direction. Anyone paying attention can feel it. But individual effort may no longer connect strongly enough to reverse the drift. That is not because effort is fake. It is because late effort enters a process that is already carrying more momentum than one person can counteract. Once again, the signs remain available after leverage has begun to fade.

Modern life amplifies this everywhere. We live in conditions of constant readout. More metrics, more feeds, more commentary, more sensors, more narrative, more public traces. In one sense, this makes the world easier to read. But the same world also makes action more diluted. Systems are larger. Decisions are spread across many people and many layers. Reactions lag. Signals arrive mixed with noise. Personal moves are absorbed into crowds, platforms, institutions, markets, and habits that are already in motion. It becomes easier to detect direction and harder to bend it locally.

This is why the leverage gap feels so intimate and so impersonal at the same time. Intimate, because you feel it in your own life, in your own timing, in your own regret. Impersonal, because the pattern is often structural. The window does not close because the universe singled you out for humiliation. It closes because information and influence obey different constraints.

That realization matters emotionally. Without it, people tend to sort every bad outcome into two boxes. Either I had no warning at all, or I failed to act when I should have. But there is a third possibility, and it may be far more common than we admit. I saw the pattern while my leverage was already degrading. I was not fully blind, and I was not fully powerful. I was inside the gap.

This does not erase responsibility. It sharpens it. It tells you where responsibility really lives. Not in fantasy control. Not in perfect hindsight. In timing, in signal quality, in the ability to distinguish visibility from reach, and in the discipline to act before a process becomes obvious to everyone. By the time a trajectory feels undeniable, it may already be expensive to alter.

That may be the hardest lesson in this whole subject. Clarity is not always a gift. Sometimes clarity is what remains after easy action is gone. Sometimes the world becomes most readable at exactly the moment it becomes least forgiving.

Once you understand that, a lot of familiar pain starts making more sense. The breakup you could feel but could not reverse. The body signal you noticed before there was a clean way to act. The market danger you saw but could not turn into a safe advantage. The job decline you sensed long before the final cut. These do not all prove the same outcome, but they do share one unsettling shape. The pattern stayed alive. The handle did not.

And that is why late clarity hurts so much. It is not merely that you know. It is that knowledge remains present while contact weakens. The future keeps speaking, even as your hand slips from the controls.

The mind does not handle that split very gracefully. It wants a cleaner story than reality usually gives. If a bad outcome arrives, the mind prefers one of two explanations. Either the future was hidden, or the person failed. Either no signal existed, or courage did not. Either the road was dark, or the driver froze. Those stories are emotionally simple. They are also often false.

A great deal of suffering comes from forcing messy timing problems into moral language. Someone says, I should have done more. Maybe. But maybe the deeper truth is that they were trying to act inside a narrowing window, in a situation where feedback was slow, noise was high, and their influence no longer traveled cleanly enough to produce a reliable change. That is a very different diagnosis of the pain. It does not turn every person into a victim of fate. It simply refuses to confuse late visibility with full power.

This matters because modern culture flatters the fantasy of control. It tells people that awareness is almost the same as mastery. If you can see the pattern, then surely you can beat it. If you understand the trend, then surely you can redirect it. If you sense a danger early, then surely the outcome still belongs to you. That sounds empowering, but it quietly loads ordinary life with impossible moral weight. It teaches people to interpret every visible disaster as a failure of personal will. It makes them feel guilty for not turning foresight into rescue.

Reality is harsher and more mechanical than that. Many processes become easy to notice precisely because they have already gathered force. They are no longer tiny. They are no longer hidden in the noise. They have grown large enough to throw shadows. The person can finally see the shadow clearly. But seeing the shadow does not shrink the thing that casts it.

Think about debt, personal or public. Early on, debt can look manageable because the numbers are still small, the payments still feel possible, the story still sounds temporary. Later, the pattern becomes much easier to read. The pressure starts appearing everywhere. Choices narrow. Stress rises. Small disruptions suddenly matter. The future becomes more legible because the process now marks every corner of life. But that same growth in clarity often comes with a collapse in flexibility. The person can see the trap more clearly than before, yet have less room to move. Legibility rises, maneuverability falls.

Addictions often produce the same shape. The pattern becomes obvious long before the grip becomes weak. In many cases, the person knows very well where things are heading. The trajectory is not a mystery. They can describe it with painful accuracy. They can tell you what tomorrow looks like, what next month looks like, what another year looks like. The future is not hidden from them. But knowledge is not identical to control. Habits have momentum. Cues have timing. Stress has gravity. The body learns loops that thought alone cannot easily interrupt. This does not make change impossible. It does show why prediction can stay alive even when action feels underpowered.

The same pattern appears at larger scales too. A city can sense that housing is becoming distorted before anyone can fix affordability. A country can sense that public trust is thinning before any clean repair becomes available. A market can sense fragility before timing, cost, and crowding make a profitable response difficult. A family can sense a long decline before any one conversation can reverse it. Again and again, the same geometry appears. A process becomes more visible as it becomes harder to steer.

That is why the leverage gap is not just a scientific curiosity. It is a human wound. It helps explain why so many people feel haunted by scenes in which they were not blind, yet still ended up overwhelmed. They did not miss the drift completely. They caught enough of it to suffer in advance. They watched the future approach with just enough clarity to feel sick, and not enough practical reach to feel safe.

There is a special cruelty in that. Pure surprise can be sharp, but it is often short. Late clarity is different. It stretches pain across time. The person lives through the bad future twice. First as anticipation, then as arrival. They feel the coming weight before the formal event lands, then they feel it again when the world finally catches up to what they already sensed. That double burden can make people feel strangely foolish. They think, if I already knew, why am I still here. But that question assumes knowledge and leverage move together. Often they do not.

In fact, one reason the leverage gap hurts so much is that visibility can produce false confidence. Seeing the direction of a process can create the feeling that intervention should still be easy. The future looks close enough to touch. The clues are vivid. The pattern feels undeniable. Surely something can still be done. Sometimes that is true. But sometimes the visibility is deceptive in a very specific way. It tells you where the train is going, while hiding how little force your hands now have on the tracks.

Medicine offers a useful analogy here. A disease can become obvious long before it becomes easy to reverse. A fracture on an image may be clear, but the healing is still slow. A tumor may be visible, but visibility is not surgery. A bad number on a test can reveal the danger, but revelation itself does not restore the body. Here again, readability and steerability are separate things. People often accept that more easily in medicine than in ordinary life. They know a clean scan is not the same thing as a cure. Yet in relationships, careers, money, and culture, they keep slipping back into the fantasy that recognition should automatically deliver power.

It does not.

This is one reason regret is such a poor historian. Regret looks backward and compresses time. It takes a long sequence of uncertainty, half-signals, hesitation, weak feedback, mixed incentives, and fading options, then rewrites all of it as one clean moment where the right move should have been obvious. It edits out the noise. It edits out the delay. It edits out the cost of acting under ambiguity. It edits out the fact that people live inside events, not above them. In hindsight, the person appears to have stood at a bright crossroads with a clear sign. In reality, they were moving through fog with a map that kept updating late.

That is why we need a more honest language for helplessness. Helplessness is not always the absence of action. Sometimes it is action inside a badly weakened channel. Sometimes it is effort that arrives after the world has become less responsive. Sometimes it is a hand on the wheel after the steering has gone soft. That does not make the effort fake. It means the connection between effort and outcome has degraded.

Once you start seeing this clearly, the emotional tone of many familiar scenes changes. The friend who stayed too long in a failing relationship may not have lacked intelligence. The worker who failed to save a collapsing role may not have lacked discipline. The trader who saw danger and still got crushed may not have lacked insight. The person who sensed illness before a formal result may not have been irrational. In each case, the relevant question is not only what they knew. The real question is what could still be moved from where they stood.

That question is harder, less flattering, and much more useful.

It forces you to care about timing instead of drama. It makes you ask whether a process is still responsive, not merely whether it is visible. It shifts attention away from heroic fantasies and toward coupling, signal quality, feedback speed, and the shrinking geometry of intervention. It asks not, do I understand the danger, but, does my action still reach it in time.

That is a major change in how a person reads life. It is also a more compassionate one. Not soft, not indulgent, just accurate. It allows you to stop treating every bad outcome as proof of weak character. Some failures really are failures of nerve. Some are failures of judgment. But some are failures of timing in a world where visibility can outlast reach.

And that may be the hardest sentence in this whole subject. Visibility can outlast reach.

Once that becomes clear, many painful memories reorganize themselves. The breakup felt obvious because the bond had already thinned enough to reveal its own future. The financial danger felt obvious because strain had already piled up into visible shape. The body felt wrong because the process had already become detectable. The public drift felt obvious because the pattern had already grown larger than any one person. In each case, the clues were real. The tragedy was not imaginary. The tragedy was that the clues arrived in a region where leverage was already fading.

This does not mean people should surrender to every dark intuition. It does not mean all visible futures are fixed. It means something narrower and more serious. There are moments when the future becomes easier to read than to change. There are moments when clarity expands while agency contracts. There are moments when the world keeps telling you the truth after it has already started withdrawing the handle.

That is the leverage gap in lived form. Not mysticism. Not doom. Not an excuse. Just one of the coldest regularities of a delayed, noisy world. The pattern remains. The grip weakens. The person feels both at once, and calls it failure, when often it is something more precise, they had contact with the trajectory after the easy leverage was already gone.

One reason this pattern feels so modern is that the world now produces visibility faster than it produces usable control. We live surrounded by signals. Notifications, indicators, headlines, metrics, dashboards, feeds, ratings, trends, alerts, models, forecasts, performance graphs, public mood readings, comment streams, search data, watchlists, health trackers, price charts. It is one long glowing wall of partial legibility. The future leaks through everything.

But there is a difference between living in a world full of readouts and living in a world where local action still matters in proportion to what you can see. Those are not the same world. We often pretend they are the same because they feel similar from the inside. Seeing more creates the emotional impression that you should also be able to do more. If the map is brighter, surely the steering must be better. If the warnings are louder, surely the intervention window must still be open. If the trend is obvious, surely somebody can still turn it.

Yet that is exactly where modern life becomes slippery. In many domains, the map has improved faster than the steering.

Take digital platforms. They are giant machines for turning human behavior into visible pattern. They track attention, reaction, hesitation, curiosity, purchase intent, outrage, boredom, drift, repetition, drop-off, return. They are built to detect direction early. They can often sense what a population is going to do before the population has language for it. They can detect the curve before people consciously experience themselves as moving along it. That is a form of legibility. But from the user's side, personal control over the larger process is weak. A person may know that a platform is changing their mood, shortening their attention, rewarding their worst impulses, amplifying conflict, or bending public life into weird shapes. They may see it clearly. They may even be correct in detail. But their individual leverage over the larger mechanism is tiny. Their local view improves faster than their local power.

This is one of the signatures of the leverage gap in the modern world. Systems become increasingly good at exposing pattern while remaining stubbornly resistant to local steering.

Finance does it too, and often with almost comic cruelty. Information is abundant. Prices move in public. Narratives spread instantly. Liquidity conditions, positioning, sentiment, volatility, policy expectations, technical fragility, all of it becomes visible in fragments. It is possible to see stress building. It is possible to feel instability before the official break. Yet the very speed and crowding of the environment mean that acting on the pattern is hard. The same transparency that helps reveal the trajectory also attracts competition, front-running, panic, overreaction, noise, and timing risk. More people can see the shape. That does not mean more people can convert that sight into clean influence. In some cases the opposite is true. The more visible the opportunity becomes, the less room there is to exploit it without cost.

This is why markets are such a useful teaching device for FAWP. They strip away comforting myths. They show, with unusual honesty, that being right is not the same as being early enough, large enough, fast enough, or solvent enough. They show that recognition can survive in conditions where intervention becomes punishingly difficult. A trader can see the cliff and still be unable to cross the terrain between insight and action.

Large institutions create a slower version of the same pain. An employee can read decline before leadership admits it. A manager can sense drift before the numbers formally collapse. A citizen can feel a public process hardening long before any visible change arrives. In these environments the signs are often there, but they are smeared across time. Nothing dramatic happens on one day. Instead the direction leaks through tone, through budget choices, through strange silences, through talent leaving, through priorities narrowing, through repeated delays that begin to rhyme. The process becomes legible through accumulation. Yet because the process is large and distributed, no single actor has clean leverage over it. Everyone can feel the ship listing. Very few people can reach the rudder. That image matters because modern systems are often built in exactly that way. They create long intervals of visible drift with weak local control. You can detect the direction before you can do much to reverse it. In fact, the scale of the thing is often what makes the direction visible. It gets large enough to produce unmistakable symptoms. But that same largeness is what makes it hard to steer from any one point inside it.

The body, strangely enough, now belongs in this story too. Not because human biology changed, but because modern self-measurement changed the feeling of living inside a body. People track sleep, heart rate, glucose, strain, recovery, movement, stress, blood markers, step counts, weight trends, resting patterns. The body has become more legible to itself through devices and repeated measurement. That can be helpful. It can genuinely improve early awareness. But it can also intensify a familiar confusion. More awareness of fluctuation does not automatically produce more control over the deep processes beneath it. A person may know more about their trajectory than any generation before them and still face the same hard truth that intervention has costs, delays, limits, and windows. More readout is not identical to more agency.

This is one of the great psychological traps of our era. We mistake monitoring for mastery.

To monitor something is to receive repeated information about it. To master something is to have reliable, timely, well-coupled influence over its trajectory. Those can overlap, but they do not always. In some cases monitoring mainly increases the emotional vividness of helplessness. You watch the line move. You watch the signals worsen. You watch the trend confirm itself. The process becomes more undeniable by the week. But the handle remains weak, expensive, late, or structurally unavailable. The tragedy is not that the person knows nothing. The tragedy is that they know enough to watch the future thicken.

Social life works this way at scale too. Public moods become legible before they become tractable. Cultural fatigue becomes visible before there is any clear intervention point. Distrust spreads through repeated small fractures until almost everyone can feel it, but no one local conversation can restore the missing fabric. A society can know something is bending in a bad direction while remaining unable to coordinate the kind of timely, coherent, high-coupling response that would shift it. The result is one of the most maddening forms of collective experience. Many people recognize the same trajectory. That shared recognition does not add up to shared power.

This is where the leverage gap stops being only personal regret and becomes a civilizational mood. People begin living with obvious trajectories that still feel unreachable. They can sense the shape of the problem, discuss it endlessly, generate articles, panel talks, documentaries, graphs, slogans, and expert warnings, and still watch the process continue. The information layer becomes richer. The action layer remains thin. That is not because nobody cares. It is because visibility and steerability are governed by different bottlenecks.

And once you see that, another strange feature of modern life becomes easier to understand. The world often feels both overexplained and undercontrolled at the same time. There is no shortage of commentary. There is no shortage of pattern recognition. There is no shortage of retrospective clarity. Yet outcomes continue to arrive with the force of inevitability. That creates a cultural atmosphere full of frustration and blame. People feel that if everything was so visible, someone must have failed. Sometimes someone did. But often the deeper truth is less theatrical. The pattern was widely legible in a region where effective leverage was already thin, delayed, crowded, or misaligned.

That is why the leverage gap should not be dismissed as an abstract concept for researchers. It is a practical description of why so many informed people still feel trapped. They are not always trapped because reality is utterly unknowable. Often they are trapped because reality becomes increasingly knowable in forms that do not automatically grant a good handle on timing, reach, or intervention.

The closer a society gets to total readout, the more tempting it becomes to believe that control should follow naturally. But a wall of instruments is not the same thing as a wheel that still turns the machine. A warning light is not a brake. A beautiful forecast is not a rescue. A diagnosis is not a cure. A market thesis is not a profit. A moment of relationship clarity is not a repaired bond. Again and again the same lesson returns. Legibility and leverage are neighbors, not twins.

That is why late modern life can feel so psychologically exhausting. People are asked to live in a condition where the future keeps showing itself in outline, but local influence does not scale to match the brightness of the outline. They see more. They control less than that degree of sight seems to promise. The result is a constant temptation toward self-blame. If the signals were there, why did I not save it. If the trend was visible, why did I not stop it. If I could describe the danger, why was I still caught inside it.

Sometimes the honest answer is painful but simple. Because the world let you read the weather before it let you touch the storm.

That sentence sounds severe, and it needs to be handled carefully, because at this point people often make the wrong leap. They hear that the future can stay visible while leverage weakens, and they rush toward extremes. Either they turn the idea into mysticism, as if ordinary life were full of private prophecies and hidden destinies, or they turn it into despair, as if action never mattered and every outcome were already sealed. Both reactions miss the point.

The point is narrower, more disciplined, and much more useful. The leverage gap does not mean you possess magical access to the future. It does not mean you can see everything. It does not mean your intuitions are always correct. It does not mean every fear is a forecast. It does not mean all visible trajectories are fixed. It certainly does not mean effort is meaningless. What it means is simpler. In many real situations, the information channel and the action channel are not equally healthy. The world may still be telling you something real while your ability to redirect it is already fading. That distinction matters because human beings are very good at flattering themselves with bad theories. A person notices a pattern once or twice, then starts treating every hunch like a deep revelation. Another person gets hurt by a process they could partly see, then decides that nothing is ever steerable and all action is theater. Both mistakes are tempting because they turn a difficult truth into a cleaner story. One story makes the self feel special. The other makes the self feel excused. Neither one helps.

A disciplined view begins by admitting how messy ordinary recognition really is. Most of the time, you do not receive a perfect signal. You receive fragments. You receive odd timings, repeated tones, strangely similar days, numbers that start leaning in one direction, moods that stop recovering, behaviors that no longer feel random, prices that keep reflecting the same hidden pressure, institutions that begin speaking in thinner language, bodies that keep returning to the same complaint. None of this arrives as a prophecy. It arrives as weak but accumulating structure. The future is not revealed to you in full. It becomes more legible by degrees.

That is why the leverage gap should never be confused with certainty. Certainty is much rarer than people think. In ordinary life, you are usually working with uneven evidence. You are estimating.

You are asking whether the present is carrying a real shape of the future or whether you are merely projecting anxiety onto noise. That is hard. It is one reason people miss windows. They are not only indecisive. They are often trying to distinguish signal from distortion under pressure, with incomplete feedback and real costs for acting too early.

This matters especially in finance, because finance attracts bad philosophy the way sugar attracts flies. People love to speak as if markets reward pure intelligence. They imagine that if you see the crash, you win. If you understand the structure, you profit. But markets are one of the best places to learn the opposite lesson. A person can be directionally right and operationally dead wrong. They can identify fragility, leverage, overextension, liquidity risk, policy mismatch, valuation absurdity, crowd psychology, and still fail to convert that recognition into a survivable position. Why. Because timing matters. Carry matters. path dependency matters. Margin matters. Volatility matters. The market does not pay you for being conceptually correct. It pays, when it pays at all, for being correctly positioned within a narrow and often brutal geometry of time and exposure.

That is exactly why the leverage gap is not an excuse for lazy hindsight. It does not permit someone to say, I knew everything, therefore my loss proves that action was impossible. No. Sometimes the person did not know enough. Sometimes their model was wrong. Sometimes they misread noise as pattern. Sometimes they acted badly. Sometimes they wanted the dignity of foresight without the discipline of execution. The leverage gap is not a blanket pardon for every failed move. It is a framework for understanding a particular structure of failure, the one where prediction remains meaningfully alive while steerability has thinned below practical strength.

The same caution applies to relationships. It is very easy to romanticize hindsight there. A person says, I always knew this would end. Maybe they did. Maybe the signs were real. But maybe they are also rewriting a long period of ambiguity into a cleaner memory than it really was. Human beings are astonishingly talented at editing their own past into a sequence of obvious warnings. That is why we have to be careful. The leverage gap does not mean every sad intuition was correct. It means there are cases where the bond really is leaking its future before the people inside it can still repair it efficiently. That difference is everything. It preserves seriousness without collapsing into superstition. It lets us say that patterns can be real without pretending they are infallible. It lets us say that late clarity hurts without turning pain into proof of destiny.

There is another mistake to avoid here, and it is almost the opposite one. Some people hear this whole idea and decide it means we should become passive. If leverage can die before visibility does, why try. Why intervene. Why plan. Why strain. Why take action under uncertainty if the window may already be closing. That reaction is understandable, but it gets the lesson exactly backward. The real lesson is not, action is pointless. The real lesson is, action is timing sensitive.

That is a far more demanding conclusion. It says that one of the great errors in human life is waiting for emotional certainty in environments where leverage expires before certainty arrives. People often hope the world will become obvious first, and only then will they move. But many important processes do not honor that wish. By the time the pattern becomes undeniable, the cost of changing it may have risen sharply. The relationship is no longer soft enough to repair easily. The body process has advanced. The market edge has crowded out. The institution has hardened its direction. The social drift has built its own reinforcing loop. What feels like procrastination in hindsight may really be a mismatch between the speed at which legibility improves and the speed at which leverage decays.

This is why the leverage gap should make a person more serious, not less serious, about early action. It should make you more respectful of weak signals, more attentive to timing, more aware that delayed clarity can be a trap. It should not make you worship every hunch. It should make you ask a better question. Not merely, do I understand what is happening. But, is this still a regime in which my action can travel far enough, fast enough, strongly enough to matter.

That is a harder question because it forces humility. It admits that understanding is not the same as command. It also admits that some environments train people to overestimate the power of awareness itself. We now live in a culture that treats observation as half of mastery. Track it, monitor it, measure it, talk about it, map it, visualize it, and surely control will follow. But that is often false. Sometimes observation mostly gives you sharper contact with your own shrinking window.

You can see this in health culture very clearly. A person can track ten different markers and still not possess a clean way to change the deeper process those markers reflect. The numbers are not useless. Awareness matters. Monitoring matters. But awareness does not dissolve biology. It does not erase delay. It does not remove constraints. It does not grant omnipotence. The body is not persuaded by dashboards.

This is why the leverage gap is not anti knowledge. It is anti confusion. It does not say that knowing is worthless. Knowing can be enormously valuable. It can help you prepare, reduce exposure, preserve options, reframe blame, choose timing better, protect what remains reachable, and avoid stupid fantasies of control. In many cases, that is a tremendous gain. The point is simply that knowledge has to be understood for what it is. Sometimes it is a steering advantage. Sometimes it is mainly a forecasting advantage. Sometimes it is both. Sometimes it is only enough to help you brace, not enough to help you redirect.

There is dignity in learning that difference. It frees you from two childish dreams at once. The first dream is that if you are smart enough, you will always be in command. The second dream is that if command is incomplete, nothing matters. Both are forms of vanity. Reality is more demanding and more interesting than either one. You are not all powerful. You are not powerless in every domain either. You are an agent moving through environments where different channels stay alive for different lengths of time.

Some of those environments still let action bite. Some mostly let you read. Some let you prepare but not steer. Some let you reduce damage without reversing direction. Some let you act only if you move before the pattern becomes comfortable to name. That is the adult picture. Uneven, unsentimental, and far more useful than the fantasy that every visible future is still equally reachable. Once you understand that, the leverage gap stops sounding like doom and starts sounding like a measurement problem inside life itself. The question is no longer whether you know everything, or whether you control everything. The real question is more exact, and much more humane. What kind of contact with the future do you still have from here.

One of the clearest places to see all of this is in the difference between warning and rescue. People often imagine that if a warning is real, rescue must still be nearby. That seems reasonable at first. If the smoke alarm is loud, the danger should still be manageable. If the body is speaking clearly, help should still be able to catch up. If the numbers look ugly, correction should still be within reach. If the relationship is obviously deteriorating, one honest conversation should still be able to save it. If the market looks ridiculous, a smart position should still be able to profit from the coming break. But the world does not always keep warning and rescue on the same timetable.

In many cases the warning gets stronger as rescue gets weaker.

That is one of the hardest truths for people to accept, because it offends the emotional logic of fairness. We want reality to reward recognition. We want the person who sees the danger to receive a meaningful advantage over the person who stays asleep. Sometimes that happens. Sometimes early recognition really does preserve options, reduce damage, and create room to move. But not always. Sometimes recognition arrives in a grim, narrow form. It tells you what is coming without restoring the kind of leverage that would make escape clean.

Think of a shoreline before a storm. The sky changes first. The air changes. The water changes. The birds change. The pattern announces itself in advance, but not in the same language as intervention. A person standing there may know perfectly well that weather is turning violent. That knowledge matters. It may tell them to leave, to secure what they can, to avoid standing in the wrong place. But it does not grant command over the sea. The warning is real, but rescue is limited to what can still be done at the edge. Preparation is available. Mastery is not.

A lot of modern life has that shape. The signals keep getting better at telling you what kind of future is approaching, but not necessarily at giving you proportional power over it. That is why so many intelligent people feel trapped in highly legible situations. They are not failing to notice. They are noticing in a world where noticing and steering no longer travel together.

This becomes especially painful when people confuse preparation with defeat. Suppose someone sees the trend early and chooses not to chase the fantasy of complete reversal. They reduce exposure. They save money. They leave before the collapse gets theatrical. They soften the landing instead of trying to perform a miracle. From the outside, that can look like weakness. People often celebrate dramatic resistance and quietly underestimate timely repositioning. But once you understand the leverage gap, a different picture appears. In many circumstances, the smartest form of agency is not reversal. It is damage control while contact still remains.

That idea is emotionally difficult because it asks people to give up some romance. It says that in a delayed world, maturity often looks less like heroic conquest and more like timely adjustment. You do not always get to turn the ship around. Sometimes the real win is noticing that the rudder has become weak, and changing your own position before the full force of the turn arrives. That is not surrender. It is accurate reading of the coupling that still exists.

Finance understands this in its bones, even when people refuse to admit it out loud. Some of the most important decisions in markets are not about domination. They are about exposure. How much risk are you carrying into a future that is starting to show its teeth. How much room do you still have to be wrong about timing. How much pain can you absorb while waiting for recognition to become consensus. Very often the smartest move is not to imagine that sight guarantees profit. It is to understand what kind of contact your position actually has with the unfolding environment. Can you still act with strength from here, or has your role already shifted from steering to surviving.

This is one reason the leverage gap can feel humiliating. It forces people to downgrade their self image at the exact moment they want to feel most powerful. They have finally recognized the pattern. They can finally describe the danger. They can finally explain what is going wrong. And yet the honest conclusion may be that the range of meaningful action is narrower than they hoped. Human pride does not enjoy that sentence. Pride wants late clarity to confer late dominance. Reality often refuses.

But there is something strangely liberating in giving up that fantasy. Once you stop demanding that every act of recognition become a miracle of control, you can start seeing the real menu of action more clearly. Sometimes you can still redirect. Sometimes you can only reduce losses. Sometimes you can protect one corner of life while losing another. Sometimes you can preserve dignity, timing, liquidity, trust, health, or attention even if you cannot fully reverse the larger drift. These are not trivial gains. They are forms of agency that become visible only when you stop worshipping total control.

This is also where the leverage gap begins to teach a deeper lesson about responsibility. Responsibility is not the same as omnipotence. A person can be responsible for reading a situation honestly, for acting within the range that remains open, for avoiding denial, for protecting what is still reachable, and for not confusing noise with destiny. That is already a demanding standard. It does not require pretending that every visible future is equally bendable.

In fact, one of the most destructive habits in modern life is the constant moral inflation of control. People are told that with enough awareness, enough grit, enough optimization, enough data, enough reflection, enough personal honesty, enough hustle, they can bend almost anything. That message sounds empowering, but it often becomes a machine for manufacturing shame. When the relationship fails, the diagnosis lands, the institution decays, the market breaks, the person looks back and thinks, I must have lacked force. Yet force is not the only variable. Timing matters. Responsiveness matters. Feedback matters. Delay matters. The thickness of the environment matters. The shape of the channel matters. Reality is not a motivational poster.

This is why the leverage gap can actually make a person kinder without making them soft. It encourages a more exact kind of judgment. Instead of asking only whether someone cared enough, believed enough, or tried hard enough, it asks what kind of contact with the future was still available to them from where they stood. That question does not erase accountability. It locates accountability in a more truthful place.

It also changes how we think about courage. Courage is often pictured as the refusal to look away from danger. That is part of it. But in a world where readability can outlast reach, courage also means accepting the scale of what can still be influenced. It means not lying to yourself about the state of the window. It means not mistaking late emotional intensity for late mechanical power. It means acting where contact remains alive, instead of burning yourself in rituals of false command.

That kind of courage is much less cinematic. It does not always produce a comeback story. Sometimes it produces a graceful exit, a reduced position, an early apology, a smaller commitment, a medical check before the evidence feels complete, a refusal to add more debt to a visibly sick process, a decision to protect attention from forces that are already shaping it. These moves can look humble, even anticlimactic. But they often represent much better contact with reality than grand declarations made after the channel has weakened.

There is another reason this matters. When people do not understand the leverage gap, they often stay too long in situations that have already become structurally hostile to rescue. They keep waiting for one more confirmation, one more signal, one more undeniable proof, one more dramatic moment that will finally justify action. But because legibility often improves as leverage decays, this waiting game is dangerous. The environment becomes easier to name while becoming harder to change. The person mistakes increasing clarity for preserved opportunity.

That is a brutal trap. It teaches the wrong lesson from the very evidence that should have helped. The more obvious the pattern becomes, the more the person assumes action can still arrive in time. Sometimes the opposite is true. Sometimes the mounting evidence is not a gift of extra time. It is the visible face of shrinking room.

Once you see that clearly, many old experiences stop looking like pure failure and start looking like timing mismatches in a world that reveals trajectories unevenly. The relationship became easier to describe after it became harder to save. The health problem became easier to detect after it became harder to reverse. The financial danger became easier to explain after it became harder to trade cleanly. The public drift became easier to discuss after it became harder to redirect from any one local point. The clues were real. The pain was real. The misunderstanding lay in assuming that stronger warning always meant stronger command. It does not.

Sometimes warning is what remains when command is already thinning. Sometimes the world grows loud precisely because it is no longer especially responsive. Sometimes the future becomes most visible when the easy part of influence has already slipped into the past.

That is not a mystical thought. It is a practical one. It means the real challenge is not merely learning to notice. It is learning to notice early enough, and honestly enough, to tell the difference between a future you can still bend, a future you can only brace for, and a future that now offers mainly the dignity of clear sight.

This is why one of the most important upgrades a person can make is learning to separate three very different experiences that often get mixed together. The first is blindness, when the pattern is still hidden from you. The second is leverage, when the pattern is visible and your action still has enough contact with the future to change its course. The third is the hardest one, the one most people do not have clean language for, when the pattern is visible, but your contact with the future has already weakened so much that control becomes partial, late, expensive, or unreliable.

That third condition is where a great deal of modern suffering lives.

People know this feeling long before they know its name. They know it when they can feel a relationship cooling, but cannot seem to reach the missing center. They know it when they can sense a job losing oxygen, but cannot restore the vanished momentum. They know it when they can feel the body giving warnings, but cannot turn awareness into immediate repair. They know it when they can read strain in a market, but cannot translate insight into safe, timely gain. They know it when they can sense a public mood changing, a platform bending behavior, a social process hardening, and still cannot find a handle large enough to move the machine.

Again and again, the same shape returns. The future does not stay hidden. It shows itself. It leaks through tone, delay, repetition, fragility, drift, timing, silence, volatility, exhaustion, accumulation. It becomes legible. But legibility is not the same thing as reach.

That is the difference this whole discussion has been trying to force into the open. Most people are taught to divide life too crudely. Either you did not know, or you knew and failed. Either the danger was invisible, or you lacked courage. Either the warning never came, or you wasted your chance. But real life is often harsher and stranger than that. There are situations where the warning is real and the chance is already thinning. There are situations where the future is easier to read than to alter. There are situations where clarity is not the opening of power, but the leftover glow after power has already started draining away.

Once you understand that, many painful scenes stop looking like private stupidity and start looking like contact with a very old structural fact. Not every visible process remains equally steerable. Not every forecastable future remains equally reachable. Not every true warning comes with a matching rescue path. These are not signs of cosmic unfairness. They are features of a world where timing, coupling, delay, and feedback matter.

This realization can feel cold at first. It seems to strip away romance. It tells you that understanding something does not place it in your hands. It tells you that the world can remain readable even while it grows less responsive. It tells you that some of the sharpest moments of recognition arrive in environments where the machinery has already built too much momentum for easy correction. That is not an easy thing to hear.

But there is another side to it, and it is an important one. Once you stop confusing visibility with control, you become much less likely to waste your life on fake versions of both. You become less likely to worship late clarity as if it were the same thing as preserved power. You become less likely to wait for perfect proof in systems that punish delay. You become less likely to mistake commentary for action, monitoring for mastery, emotional intensity for leverage, or endless awareness for a real handle on outcomes. In other words, you become more honest.

That honesty is not small. It changes how a person moves through the world. It teaches you to ask better questions. Not merely, what do I think will happen. Not merely, am I seeing this clearly. But, what kind of contact with the future still exists from here. Is this still a situation where small action matters. Is this a place where speed matters more than certainty. Is this a place where feedback is still good enough to learn quickly. Is this a place where the window is narrowing faster than my confidence is rising. Is this a place where the intelligent move is to push harder, or to protect what remains reachable while the larger process continues.

Those are harder questions than the ones people usually ask. They are less flattering. They do not make you feel like the hero of a motivational poster. But they are far more useful. They force you to care about timing, scale, signal quality, and the actual geometry of influence. They remind you that the central problem in life is not always ignorance. Often it is late contact with a process that has already gathered force.

That may be one of the defining experiences of the modern world. Not simple ignorance, but visible trajectories. Not total darkness, but partial illumination without matching grip. We live amid systems that reveal themselves constantly. Markets broadcast stress. Bodies leak information. Relationships signal erosion. Institutions display drift. Platforms expose patterns. Societies produce warnings, graphs, commentary, and mood indicators almost without pause. The future leaves marks everywhere. Yet those same environments often thin out local control. The person sees more, and assumes they should therefore be able to do more. Then reality refuses. The map brightens. The steering softens. And the person calls that failure, when it is often something more exact, they were standing in the leverage gap.

That phrase matters because it gives structure to an otherwise shapeless pain. It tells you that the agony was not always madness, not always weakness, not always laziness, not always bad character, not always a lack of care. Sometimes it was a mismatch between two channels that people keep pretending are the same. The information channel remained alive. The action channel degraded. The future kept speaking, but your influence no longer traveled with equal force.

That is the cruel center of the matter. Sometimes the world becomes clearest at the exact moment it becomes less willing to be moved.

If that sounds severe, it is because it should. It is severe. It asks people to let go of one of their favorite illusions, the belief that once a pattern becomes visible, power is naturally close behind. But letting go of that illusion also gives something back. It gives precision. It gives a cleaner reading of regret. It gives a more truthful picture of responsibility. It gives a sharper sense of when action matters most, which is often earlier than comfort wants and earlier than certainty allows. It gives a more mature understanding of freedom, not as abstract unlimited choice, but as usable contact with reality before the window closes too far.

And perhaps most importantly, it gives a person a way to stop turning every late moment of recognition into a moral insult against themselves. Some failures really do come from poor judgment. Some come from fear. Some come from denial. But some come from being inside a world where pattern visibility outlives practical control. Some come from seeing the road after the steering has already gone soft. Some come from understanding a danger in a regime where the machine has become more readable than reachable.

There is no comfort prize hidden inside that truth. It does not flatter anyone. But it does clarify something that matters. It explains why some of the most painful moments in life are not the ones where nothing could be seen, but the ones where something could be seen and still not cleanly changed. It explains why late clarity can feel heavier than surprise. Surprise wounds once. Late clarity wounds twice, first as recognition, then as arrival. So this is the final shape of the idea.

The future can remain legible after leverage has begun to die.

A pattern can stay visible after your hand has lost much of its force on the wheel.

A warning can grow louder while the channel for rescue grows thinner.

The map can stay clear after the controls start failing.

That is not prophecy. It is not doom. It is not an excuse. It is a structural fact about life in delayed, noisy, high friction environments.

And once you learn to see it, a great many ordinary experiences stop looking random.

The breakup, the layoff, the diagnosis, the market break, the social drift, the institutional decay, the private collapse, the public mood change, all of them begin to share a family resemblance. They are not identical events, but they can all carry the same hidden geometry. First the process gathers momentum. Then the process leaves traces. Then the traces become easier to read. Then the person mistakes readability for preserved control. Then the outcome arrives with the sickening force of something that felt both obvious and unreachable.

That is the leverage gap.

It is one of the coldest patterns in modern life, and one of the most important to understand. Because some of the clearest moments you will ever have are the ones in which your power has already started to disappear.

Chapter 5

A real idea has to survive contact with measurement. If Future Access Without Presence were only a striking phrase for a familiar feeling, it would still be emotionally useful, but it would remain incomplete. People do know the feeling. They know what it is like to sense where things are heading while lacking enough reach to stop the outcome. They know the pain of seeing the slope before they can still change the direction. But if that experience reflects something structural, then it should not live only in memory, confession, regret, or metaphor. It should leave a mark that can be tracked.

This is where many ideas become weaker than they first appear. They sound powerful in ordinary language because they fit the texture of life so well. They feel true. They seem to name something that had always been there without a proper name. But feeling true is not the same as being tested. A person can recognize themselves in a sentence and still have no way to tell whether the sentence describes a real pattern in the world or only a persuasive style of speaking. That is why this part matters. It is the point where the book has to risk becoming concrete.

The claim up to now has been simple in form, even if its implications are deep. The present can sometimes tell you where the future is going, even after your practical ability to alter that future has begun to fade. If that claim is serious, then the two sides of it have to be pulled apart and measured separately. One side concerns legibility. How much does the present reveal about what comes next. The other concerns leverage. How much can an action taken now still shape what comes next. If those two capacities really come apart, then they should not merely feel different. They should behave differently.

That is the crucial move. Future Access Without Presence cannot remain a mood. It has to become an operational question.

Does predictive information remain detectable after steering influence falls below detection. Does the future stay readable after the action channel weakens past practical reach. If the answer is yes, then the idea stands on firmer ground. If the answer is no, then the whole framework has to shrink.

This is why measurement matters here more than usual. The subject invites exaggeration, and exaggeration is cheap. It is easy to talk loosely about sensing, knowing, foreseeing, or feeling the curve of events. It is much harder to ask what exactly survives into the present, how strongly it survives, and whether action still has enough contact with the outcome to matter. But that harder question is the one that decides whether we are dealing with a real structure or only a compelling description of regret. So the task was clear. The feeling had to be translated into something the world could answer back to.

That translation began with a simple decision. The problem could not be measured as one thing, because it was never one thing to begin with. The whole force of Future Access Without Presence comes from a split. On one side is prediction, the degree to which the present contains usable information about what happens later. On the other side is steering, the degree to which something done now still changes what happens later. If those two capacities are treated as though they are identical, the entire phenomenon disappears before it can even be tested.

This distinction sounds obvious once stated, but most people do not naturally think this way. In ordinary life, knowing and influencing are constantly blurred together. A person assumes that if they can see what is coming, they should still be able to do something about it. That assumption works well enough in small, immediate settings where signals are clean and response is fast. It works far less well in delayed, noisy, distributed environments where a process can remain legible even after the practical handle has begun to slip away. That is exactly why the two channels had to be separated.

The prediction channel asks one question. How much does the present really tell us about the future beyond chance. Not in a poetic sense, and not in a vague emotional sense, but in an operational one. If you look at what is available now, how much usable information is there about what happens next. This is the legibility side of the story. It is about whether structure remains visible.

The steering channel asks a different question. If an action is taken now, how much detectable influence does that action still have on what follows. This is the leverage side of the story. It is about whether the system still answers the hand that reaches toward it. You can think of it as the difference between reading the road and still being able to turn the wheel.

Once those two questions are placed side by side, the shape of the test becomes much cleaner. Future Access Without Presence would show up not as a single dramatic effect, but as a separation between these channels. Prediction would remain present, while steering would weaken more quickly under the same conditions. The future would stay partly readable after the action path had already thinned.

But there was one more ingredient that had to be built in from the start, delay. Delay is what turns this from a philosophical contrast into a measurable pattern. If everything happened at once, the difference between prediction and steering would often be hidden inside immediate contact. The interesting question is what happens as the gap between now and later grows. Does predictive information decay at the same rate as influence. Or do they separate. Does the world remain legible farther out than it remains steerable.

And because weak effects are easy to exaggerate, both channels also had to be judged against baselines. It is not enough to see a curve and declare victory. Some apparent structure is nothing more than noise wearing a convincing shape. So the design had to ask not only whether prediction and steering were detectable, but whether they were detectable above what chance, drift, and loose interpretation could produce on their own.

That was the measurement problem in its cleanest form. Two channels, one for legibility and one for leverage, tracked across delay and checked against background. Only then could the feeling be turned into something precise enough for the world to either confirm or refuse.

The basic design followed naturally from that split. If the question is whether prediction can survive after leverage begins to die, then the only honest way to test it is to place both under the same conditions and watch how they change as those conditions become harder. That meant taking a system, introducing delay, and then asking two questions again and again at different distances from the present. How much does what is available now still tell us about what comes next. And how much does what is done now still change what comes next.

That is the whole architecture in plain form. Start close to the present, where contact is strongest. Then move farther out. At each step, measure the prediction channel and the steering channel separately. The first asks whether the future remains legible. The second asks whether the future remains touchable. If both shrink together at the same pace, then the gap may be more apparent than real. But if steering drops away faster while prediction remains detectably alive, then the pattern has a shape that deserves its own name.

Delay is what makes this visible. Close to the moment of action, many systems still allow both reading and influence. The observer can see enough, and the intervention still reaches the outcome strongly enough to matter. But as delay increases, the field changes. Noise accumulates. Intermediate processes intervene. Feedback arrives later. Small actions become diluted. Paths branch, reinforce, and harden. The result is that the future may remain structured enough to be partly forecast, even while the local action channel grows too weak to register clearly against background variation. That is the regime the test was built to find, if it existed at all.

So the procedure was not complicated in spirit, even if the implementation required care. Choose a range of delays. At each delay, ask how much present information predicts the later state. At that same delay, ask how much present action still influences the later state. Then compare both results to calibrated background levels rather than trusting raw values. This matters because weak patterns can look impressive when viewed in isolation. The eye loves a curve. The mind loves a story. But the world is full of accidental structure. A real result has to rise above what loose alignment and random fluctuation would produce on their own.

What makes the design powerful is that it does not depend on mood or interpretation. It does not ask whether the observer felt that the future was readable. It asks whether readable structure was actually present. It does not ask whether action seemed important. It asks whether action still left a detectable mark. In that sense the design is almost severe. It strips away the flattering parts of the intuition and leaves only what can survive repeated comparison.

And the outcome to watch for was very specific. Not just a gradual weakening of everything, but a separation. A point where steering becomes difficult to distinguish from background while prediction remains measurably above it. That is the core signature. A future still readable, but no longer easily moved. If such a pattern appeared consistently across delay, then Future Access Without Presence would stop being only a description of how life feels. It would become something far more serious, a measurable difference between two capacities that people had spent too long treating as one. To make that measurable in a clean way, the book needed one technical idea, but only one, and it needed it in a form a normal reader could actually carry. That idea is mutual information. The name sounds heavier than it is. In plain language, mutual information asks a very simple question. How much does knowing one thing reduce your uncertainty about another. Or more casually, how much does this really tell us about that beyond chance. Not how poetic the connection feels. Not whether two lines on a graph look similar. Not whether a story can be told after the fact. How much real dependence is there.

That made it useful for both channels at once. For prediction, mutual information asks how much the present state tells us about a later outcome. For steering, it asks how much the action taken now still tells us about what happens later. Same general logic, two different kinds of relationship. One measures legibility. The other measures leverage.

This matters because ordinary correlation is often too narrow for the kind of structure we care about here. Some relationships are not simple straight lines. Some are curved, thresholded, tangled, or mixed with changing scales. Mutual information is a more general way of asking whether a connection is really there. It does not care whether the dependency is neat and linear. It asks whether one variable still carries usable knowledge about another.

A good everyday analogy is a weather forecast. Suppose I tell you only that it is July. That gives you a little information about tomorrow's temperature, but not much. Suppose I also tell you the current temperature, pressure, wind, and humidity. Now you know much more. The present contains structure about what comes next. Mutual information is a way of quantifying that reduction in uncertainty. It is a way of asking how much the present is actually narrowing the space of possible futures.

The same logic works for action. If turning the wheel now still shapes where the car is a moment later, then present action contains information about later position. If the steering system becomes delayed, weak, noisy, or disconnected, that information begins to fall. At some point your movement of the wheel may still matter in theory, but not strongly enough to stand out from the rest of the motion. The steering channel loses detectable grip.

That is why mutual information fit the problem so well. It gave one common language for both sides of the split. It let the book ask, in the same mathematical spirit, how much the present predicts and how much the present still influences. Then the real question became visible. Do those two forms of connection decay together, or do they separate.

For the reader, the important thing is not the formula. The important thing is the meaning. Mutual information is just a disciplined way of asking whether there is still a real relationship there, above chance, above wishful interpretation, above the human tendency to see patterns because we want the story to hold. It is the simplest serious bridge between the feeling this book began with and the measurements the argument required.

Weak signals are dangerous for one simple reason, they are easy to fake by accident. The weaker the effect you are looking for, the easier it becomes to confuse a real trace with noise, drift, coincidence, or a measurement trick that flatters the eye. This is where many exciting ideas collapse. Not because nothing is there, but because a tiny pattern can look impressive long before it has earned your trust.

That is why raw curves were never enough. A line on a graph can be suggestive, but suggestion is cheap. If you want to say that prediction remains alive after steering has fallen away, then you have to show more than an appealing shape. You have to show that the pattern stands above the kind of structure that appears when alignment is broken, timing is scrambled, or chance is allowed to wear a convincing mask.

This is the job of null controls. In plain language, a null control asks what the measurement looks like when the meaningful relationship has been deliberately disrupted. If the apparent effect survives even after the relevant link has been broken, then the effect was probably never doing the work you thought it was doing. It may have been background regularity, estimator bias, random fluctuation, or some broader drift that makes weak patterns appear where no true channel remains.

That is why shuffling matters. If you randomly reorder the relevant pairings, real dependence should collapse. And that is why shifting matters. If you preserve some broad structure while breaking the timing that carried the true link, a genuine effect should weaken toward background. These are not decorative tests. They are the difference between seeing a trace and merely hoping one is there.

Calibration matters for the same reason. A weak effect should not be accepted just because it is greater than zero in some loose abstract sense. In real data, tiny values can appear for all sorts of dull reasons. So the question becomes more demanding. Not is there a number, but is the number meaningfully above the background produced by the null. That is where thresholds come in. They give the measurement a floor beneath which apparent signal is treated as operationally indistinguishable from noise.

This may sound severe, but it is exactly the kind of severity the idea needed. Future Access Without Presence was always about a narrow difference between two channels. Narrow differences are where bad methods do the most damage. Without controls, almost anything can be made to look suggestive. With controls, the field gets colder, but much clearer.

That clarity is what makes the result worth anything. If prediction remains above calibrated background while steering falls to it, then the separation is no longer just a story told after the fact. It becomes a measured distinction between what the present can still reveal and what the present can still change. What mattered most in the end was not the existence of some vague pattern, but the shape of the separation. That was the real test. If Future Access Without

Presence was only a poetic way of describing regret, then the curves should have wandered together or collapsed together. Prediction and steering might both weaken with delay, but they would not open into a clear and durable gap. The whole idea would shrink back into mood.

That is not what happened.

What appeared instead was a structure with three distinct regions. Near the present, both channels were alive. The future was readable, and action still had enough grip to leave a detectable mark. This is the ordinary zone of agency, the region where knowing and steering still overlap strongly enough that people tend to confuse them. You can see what is coming, and your hand still matters. In daily life this is the world of quick correction, small loops, and direct response. It is the part of reality most flattering to our instincts.

Then the first break appeared.

As delay increased, the steering channel began to collapse faster than the prediction channel. Not eventually in some vague sense, but measurably. After the trivial zero point, there was a short horizon beyond which steering fell to the level of calibrated background. In the flagship result, that post zero steering horizon appeared at about four delay steps. Beyond that point, present action was no longer distinguishable from the null floor in a reliable operational sense. The hand was still moving, but the future was no longer answering clearly enough for that movement to stand out.

That alone would have been interesting. But it was not the central result.

The central result was what happened next.

The process did not immediately become chaotic or unreadable once steering dropped below detection. The future did not go dark at the same moment the action channel weakened. Prediction remained alive. It remained above calibrated background not just for a moment, but across a long interval after steering had already lost practical grip. In the flagship run, there was a broad region between the post zero steering horizon and the later failure cliff, a region stretching from about four steps out to roughly thirty five. Inside that region, the future stayed legible while leverage had already become operationally absent.

That is the leverage gap in its clearest measured form.

You can think of it as a valley between two endings. The first ending is the end of reliable steering. The second is the end of even practical stability, when failure finally becomes overwhelming. Between them lies the strange middle territory this book has been describing from the beginning. The future is not hidden there. It is visible. But it is visible from a position of weakened reach.

This is why the result matters so much. It turns a familiar human experience into a measurable pattern. There really can be a region in which the present still carries information about what comes next while present action no longer carries enough influence to change what comes next in a detectable way. The world stays readable after it has already started becoming hard to move.

That is the payoff.

It is also why the result feels so recognizable once you see it. The pattern is scientific, but the emotional shape is ordinary. It looks like the relationship that is already ending before the conversation happens. It looks like the body that is already warning before the diagnosis arrives. It looks like the job that is already weakening before the notice appears. It looks like the market that is already revealing its direction after the cheap edge is gone. The measured gap does not replace lived experience. It explains why that experience has such a particular sting. Clarity survives the loss of easy control.

The later part of the result makes this even sharper. There was not just an early steering horizon. There was also a later failure cliff, around delay thirty five in the flagship case, where breakdown became overwhelming. By that point the arrangement was no longer merely hard to steer. It was functionally gone. Yet even near that cliff, predictive information still remained above null. Not at the same strength as before, of course, but still detectably present. The future continued to cast a shadow forward even as practical influence had already thinned out long before.

That combination is what gives the result its force. If prediction had vanished the moment steering did, there would be no special pattern here, only uniform decay. If steering had remained strong as long as prediction remained strong, there would be no split, only a more ordinary picture of delayed weakening. What made the result distinctive was the durable interval in which prediction outlived leverage.

This is also why the controls mattered so much. The gap did not appear as a loose visual impression. It survived calibration against null baselines. It was not recreated by the background in the same way. It had a shape, a threshold, and a persistence that random drift alone did not explain. That does not mean the result should be treated like dogma. It means it earned the right to be taken seriously.

There is a quiet philosophical weight to that. For a long time people have assumed that foresight and power rise and fall together. If you still know, you should still be able to act. If you can no longer act, then whatever you think you know must be illusion or empty anxiety. The flagship result says that assumption is too crude. There are conditions under which the future remains partly inferable after the action channel has already dropped below practical detectability.

That does not mean nothing matters. It does not mean action is always futile. It means timing matters much more than people usually admit. The gap does not erase agency, it places agency inside a narrower window than wishful thinking prefers.

And once that is seen, the entire argument of the book tightens Future Access Without Presence is not merely a dramatic phrase for feeling stuck. It is a measurable separation between two channels that human beings constantly blur together. The future can stay legible after leverage has already begun to die. That is not just how life sometimes feels. Under the right conditions, it is how the structure actually behaves.

A serious measurement has to say not only what was found, but what would have counted against it. Otherwise the whole exercise turns into branding. It becomes a way of dressing an intuition in numbers without ever giving reality the power to answer back. The point of this chapter was the opposite. If the pattern was real, it had to survive conditions under which it could also have failed.

It could have failed in a very simple way. Prediction and steering might have declined together with no durable separation. In that case the whole idea of a leverage gap would have been overstated. It also could have failed if the apparent gap were reproduced just as easily by null controls, because then the effect would look less like a real distinction and more like background structure wearing a dramatic shape. And it could have failed if the result vanished the moment calibration became strict, because then the original pattern would have been too dependent on loose thresholds or flattering interpretation.

Those possibilities matter because they remind the reader that the goal here was never to protect the theory from embarrassment. It was to expose it to conditions that could have reduced it. The fact that a durable separation remained is what makes the result meaningful.

That leaves the chapter with a clean conclusion. Future Access Without Presence is not only a human feeling, and it is not only a philosophical contrast. It can be operationalized. It can be tested. And under the right conditions, it can be detected as a measurable difference between what the present still reveals and what the present still has the power to change.

Chapter 6

There are few territories in thought that get messy faster than this one.

Latent Environmental Residual Inference (LERI) The moment people begin talking about weak access, strange anticipation, or the feeling of knowing before the fact, language starts to rot. Precision vanishes. Categories collapse. A person notices a pattern, then a story forms around it, then the story hardens into a grand claim that carries much more weight than the original signal ever deserved. By the end of the process, something small, local, and interesting has usually been inflated into something mystical, dramatic, and almost useless.

That is why this part has to begin with restraint. Not because the subject is empty, and not because every unusual experience is nonsense, but because the human mind is dangerously good at making weak evidence sound larger than it is. We are pattern seeking creatures. We are memory editing creatures. We are meaning hungry creatures. Give us three coincidences and we start building destiny. Give us one repeated feeling and we start calling it certainty. Give us a dim trace and we often rush to turn it into revelation.

Anyone who wants to think clearly in this territory has to begin there, with that uncomfortable fact about the species itself. Human beings are excellent at detecting structure, but we are also excellent at overfitting structure. We notice repetitions that matter, and we notice repetitions that do not. We detect real regularities, and we invent false ones. We remember the hit, forget the miss, and then retell the whole thing as though the pattern had been obvious from the start. It is one of the oldest habits of the mind. See a shape in fog, then swear the shape was always there.

A person dreams of an old friend, then the friend calls the next day. A trader feels that a market is unstable, then the drop finally comes. A patient says something felt wrong months before the diagnosis. A parent senses trouble before the phone rings. A culture fills itself with stories like these because stories like these are emotionally irresistible. They compress fear, memory, recognition, and chance into one dramatic package. They make life feel alive with hidden links. They flatter intuition. They flatter hindsight. They flatter the deep wish that the world might be speaking to us in a more direct voice than ordinary evidence ever does.

But this is exactly where things become sloppy. Not because the original experiences are fake, but because the interpretation often outruns the evidence. A feeling may have been real. A pattern may have been real. A weak signal may really have been there. Yet the mind takes a much larger leap. It moves from there was information somewhere in the scene, to I had direct access to what had not yet happened. Those are not the same claim. In fact, they are separated by an enormous distance.

This book cannot afford to blur that distance. If it does, the whole argument collapses into mood, suggestion, and metaphysical theater. The point here is not to make life sound enchanted. The point is to make a difficult class of experiences smaller, cleaner, and more testable. That means refusing to let vague anticipation drift upward into prophecy. It means refusing to let correlation turn into miracle. It means refusing to let every intuition borrow the prestige of the few that later lined up with events.

The first discipline, then, is to separate experience from explanation. People often have experiences that feel ahead of the event. That much is ordinary. The explanation is the dangerous part. Why did the experience occur. What information was actually present. What traces were available. What pattern was already unfolding in weak form. What background process was already shaping the field. What local signals had accumulated below the threshold of explicit articulation. These are better questions than the dramatic one. Did I somehow reach beyond the present.

Most of the time, when people use dramatic language in this territory, they are describing something more ordinary and more interesting at the same time. They are describing incomplete perception under uncertainty. They are describing local information that never rose to full conscious clarity until later. They are describing the body noticing a pattern before language catches up. They are describing memory revising itself after the outcome. They are describing coincidence made vivid by emotion. They are describing traces, drift, inertia, repetition, partial signals, and hidden regularities inside the present moment. In other words, they are often describing inference, not visitation.

That distinction matters because inference is compatible with the entire framework of this book, while visitation is not needed at all. If present accessible reality already contains structured hints, delayed consequences, residual marks, and slow moving constraints, then a future outcome can become partly inferable without requiring any magical traffic from tomorrow into today. A wave can be sensed before it breaks because the water was already moving. A crack can be felt before the wall gives way because the wall was already under strain. A bond can be sensed as failing before the breakup because the emotional weather had already changed. Nothing supernatural is required. The present was already carrying the beginning of the future in weak form.

This may sound less exciting than the mystical version, but it is actually much stronger. The mystical version explains too much and therefore explains nothing. It has no clear limits. It has no stress test. It does not tell us when the effect should weaken, when it should disappear, or why some signals should matter more than others. It gives the mind endless room to exaggerate. A disciplined inference view does the opposite. It makes the claim narrow. It says that any weak access worth discussing must ride on something local, physical, filtered, partial, and degraded. It must arrive through a real path, not through rhetorical fog.

That is why this section has to be severe with language. Words like intuition, hunch, knowing, sense, and premonition can all describe genuine human experiences, but they are dangerously loose if left unattended. They make different things sound alike. A well trained sensitivity to weak evidence sounds similar to anxious anticipation. A lucky guess sounds similar to refined pattern detection. Selective memory sounds similar to foresight. Emotional dread sounds similar to early recognition. Once all of that gets poured into the same verbal bucket, real thought becomes impossible. Everything begins to glow with false depth.

The better move is to slow down and ask what kind of event we are dealing with. Was there repeated local evidence. Was there inertia already in motion. Was there a weak but real chain from what was present to what later occurred. Was the apparent knowing really just retrospective editing, in which the mind keeps the hits and quietly buries the misses. Was the feeling generic enough that any later event could have been recruited to justify it. Was the environment already full of cues that the body registered before explicit thought did. These questions do not kill wonder. They rescue the subject from cheap wonder.

There is also a moral reason to be strict here. Sloppy thinking in this territory does not only create bad theory. It can create bad life decisions. People who romanticize weak signals can become gullible, obsessive, or frightened by noise. They can begin treating ordinary coincidence as command. They can start trusting vividness over evidence. They can become less able to distinguish between a real pattern and an emotionally charged illusion. That is not depth. That is degradation of judgment dressed up as sensitivity.

At the same time, swinging to the opposite extreme is no better. Some people hear any caution against mysticism and decide the whole territory must be empty. That is lazy too. Human beings really do pick up weak structure before they can always explain it. Bodies do register drift before language does. Environments do leak information unevenly. Outcomes do cast shadows in the present, not because the future is broadcasting itself like a radio station, but because trajectories are often already underway before they become undeniable. The mistake is not in taking weak access seriously. The mistake is in making the claim too large, too vague, and too proud.

So this is the ground rule for everything that follows. We are not interested in inflating strange experiences into cosmic drama. We are interested in shrinking them until they can be handled. If there is anything here worth preserving, it will survive reduction. It will survive loss of romance. It will survive the demand for a local path, a real trace, a filtered channel, and a clear limit. If it cannot survive that, then it was never a serious idea in the first place.

That is the right posture at the entrance to this territory. Not ridicule, and not worship. Discipline. A refusal to confuse coincidence with structure, structure with access, access with certainty, or certainty with control. Only then can we ask the question that actually matters. If weak access exists at all, what is it really made of, and how little of it can survive the journey into a human mind before noise, distance, and bad storytelling swallow the signal whole.

One of the most important mistakes people make in this territory is so simple that it almost hides in plain sight. They assume that if something exists, then it must in some meaningful sense be available. They assume that if reality contains structure, then the observer must be able to read that structure. They assume that if a pattern is there, then the mind should be able to touch it clearly enough to use it. But existence and access are not the same thing. In fact, much of human life becomes easier to understand the moment those two ideas are pulled apart.

The universe is full of things that are real without being available to you. Light from distant stars may exist, but clouds, dust, distance, timing, and the weakness of your own instruments determine whether that light ever reaches your eye in a usable form. A room may contain a faint sound, but if the wall is thick enough, the air noisy enough, or your hearing poor enough, that sound will never become part of your decision making. A city may be full of hidden motion, trains moving, cables humming, water flowing, people changing direction by the second, yet the ordinary person standing on one street corner receives only a tiny slice of that total field. The larger structure exists. The local observer only gets a narrow sample.

That is not a special problem. It is the normal condition of being embedded inside reality instead of floating above it. No human being gets the whole picture. No human being receives raw existence in full detail. We live through channels. We live through filtered arrival. We live through delayed signals, damaged traces, narrowed windows, weak cues, and incomplete views. The world may be saturated with structure, but that does not mean the structure enters the mind intact.

This is one reason so much bad thinking happens when people start talking about unusual anticipation. They jump directly from the idea that a pattern existed somewhere to the claim that they directly accessed the pattern as such. But the whole interesting question lies in the middle. What path connected the structure to the observer. What was the channel. What survived the trip. What got lost on the way. What was available in weak form, and what remained completely out of reach.

That middle layer is where disciplined thought begins. If reality contains more order than we can ordinarily articulate, that does not automatically produce mystery in the dramatic sense. It may simply mean that many processes are already underway before we consciously name them. It may mean that environments leak traces before they deliver verdicts. It may mean that the body, the senses, and memory all pick up fragments of change that never become clear enough for immediate explanation. The larger structure exists. Access arrives through scraps.

Think about how often this happens in ordinary life. A person walks into a room and immediately feels tension. They may not yet know why. No one has announced anything. No single sentence explains it. And yet the room is not neutral. Tone, posture, silence, pacing, facial stiffness, missing warmth, all of these things are already present. The pattern exists before the person can summarize it. Their access is partial. They are not outside the scene with a perfect chart of emotional variables. They are inside the scene, receiving fragments. Those fragments may still be enough to generate a real sense of direction.

The same thing happens with weather. A person can feel a storm before the storm breaks. Pressure changes, light shifts, air thickens, sound behaves differently. The atmosphere carries early structure. The person does not need magical contact with the future. They are already inside the first physical signs of what is coming. The event has not fully arrived, but the local world has begun to reorganize around it. Again, the important distinction is not between ignorance and prophecy. It is between full existence and partial access.

Health offers another simple case. The body can already be changing while conscious interpretation remains poor. Fatigue deepens, breath changes, appetite drifts, recovery slows, mood flattens. The person may not have a diagnosis, but that does not mean the process is absent. The process exists. The access path is weak, noisy, and emotionally easy to dismiss. Much of what people call intuition in health is really this. The system is already changing, and the person is receiving filtered fragments before the change can be cleanly named.

Once you see this, a large part of the mystery falls away without becoming dull. The world does not need to be less interesting. It only needs to be described more accurately. The observer is not a god, and the observer is not a blank object either. The observer is a finite readout device inside a vast structured field. That field contains more than the observer can use. Some of it arrives late. Some of it arrives weak. Some of it arrives in such damaged form that it becomes useless. Some of it arrives just strong enough to alter feeling before it can support explanation.

This matters enormously for the broader argument of this book because Future Access Without Presence depends on exactly this kind of distinction. A future outcome can be partly inferable without being directly available. The observer may receive present accessible traces that correlate with what is coming, even while lacking the leverage to alter it. In other words, the world can become legible in pieces before it becomes steerable in practice. But that legibility is never the same as full possession of the truth. It is not total access. It is partial contact.

Partial contact is the phrase worth holding onto here. Human beings live by partial contact. We do not inhabit pure darkness, and we do not inhabit perfect light. We inhabit filtered reality. That means many things are both true and unavailable at the same time. A pattern may be present without being readable. A signal may exist without being strong enough to matter. A trajectory may already be forming without the observer having enough local evidence to act comfortably. This is not a flaw in the theory. It is the starting condition of any serious theory of embedded life.

It also helps explain why the demand for certainty is so often destructive. People often want access to arrive in a form that feels complete, stable, and undeniable. They want the pattern not merely to exist, but to cross the threshold into language so clearly that no self doubt remains. But access rarely behaves that way. It often begins below that threshold. It begins as uneven pressure, weak trace, faint asymmetry, small repetition, background drift. The structure is real, but the local readout is poor. That is why a person may sense something before they can defend the sense in public, or even to themselves.

In this light, many human errors look different. Some errors come from false patterns, of course. We imagine shapes that are not there. But many other errors come from overestimating access. We assume that because reality is structured, the structure should already be available in a form we can use. We assume that if a danger exists, we should feel it clearly. If a trajectory exists, we should be able to describe it. If a signal exists, we should know what to do with it. But no such guarantee exists. Reality may be orderly while remaining operationally inaccessible at the level that matters to action.

That phrase, operationally inaccessible, is crucial. It means something can be there without being reachable in time, in clarity, or in usable detail. A locked room may contain exactly the information you need, and yet for you that information is functionally absent if you cannot open the door. A weak signal may be physically present, and yet for your purposes it may as well not exist if it arrives below the threshold of detection. The world does not owe the observer convenience. It only offers what survives the channel.

This is the deeper reason existence is not access. Existence belongs to the larger structure of reality. Access belongs to the local conditions under which an observer receives, preserves, and interprets traces from that structure. The two overlap, but never perfectly. That imperfect overlap is where human life happens. It is where uncertainty lives. It is where anticipation becomes possible without becoming magical. It is where a person can feel the outline of what is coming while still lacking clean possession of it.

And that is the exact foundation this part needs before anything more controversial can be said. If the reader does not understand that the universe can contain more order than any observer can locally use, then every later discussion of weak access will either sound mystical or absurd. But once the distinction is clear, the path opens. Reality may contain structure, yes. The real question is what fraction of that structure survives the journey into a finite observer, and in what ruined, delayed, weakened form it arrives. So what does LERI actually mean, once all the fog has been cleared away. It means something much smaller, much stricter, and much more physical than the dramatic versions people usually imagine.

It does not mean a person reaches out of the present and grabs tomorrow. It does not mean the mind escapes its own location. It does not mean a hidden gift, a private miracle, or a secret tunnel through time. It means that locally available traces, traces already present in the world around the observer, may still contain weak structure that points beyond what the observer can yet state clearly. If anything worth preserving exists in this territory, that is the form it has to take.

LERI is about residual information. The key word is residual. Not total information, not complete access, not a clean window into some distant fact, but leftover structure, faint imprint, partial carryover, the kind of information that remains after the world has already filtered, damaged, diluted, and delayed it. Think of the slight warmth left in a chair after someone stands up. Think of footprints in wet soil. Think of the smell of rain before the storm fully breaks. Think of a vibration running through a building before the machinery that caused it comes into view. In each case, the object of interest is not directly present to you in full. What is present is a local consequence, a residue, a small carrier of structure.

That is the basic intuition. A process may be underway, a trajectory may already be forming, a constraint may already be shaping what can happen next, and the local world may therefore contain weak marks of that fact before the larger picture becomes obvious. LERI is simply the disciplined claim that such marks can, in principle, be informative. Not always, not clearly, not magically, not without noise, but sometimes enough to matter.

The next word is inference. That word matters just as much as residual. LERI is not direct possession of hidden truth. It is inference from traces. This is a huge difference. Direct possession would mean the observer somehow has the distant fact itself. Inference means the observer only has a local clue and must reason outward from it, often under severe uncertainty. The difference between the two is the difference between seeing the fire and smelling smoke. The smell may be real, useful, and urgent, but it is not the fire itself. It is a degraded sign carried into the present by a real path.

That real path is the whole point. LERI only makes sense if there is some actual route by which structure reaches the observer. A local mark, a weak correlation, a surviving imprint, some present accessible carrier that has not yet been reduced to pure noise. Without that, the whole idea collapses into theater. If there is no path, then there is no readout. If there is no readout, then there is nothing to infer from. At that point the claim has become empty.

This is why LERI is not a license for grand language. In fact it should force the opposite. The moment somebody begins speaking as though they have direct contact with distant facts, the concept has already been abandoned. LERI is intentionally humble. It assumes the observer is finite, local, and stuck with damaged information. It assumes the world does not hand over full explanations. It assumes that whatever access exists must come through a narrow and lossy channel. In other words, it treats the observer as what the observer really is, a limited creature reading a large world through partial contact.

Once that is clear, a lot of examples begin to look different. Suppose a person senses that a relationship is ending before the breakup happens. The dramatic version says they somehow knew the future. The stricter version says the bond was already changing, and local traces of that change were already present. Tone had shifted. Timing had shifted. warmth had shifted. The future event had not yet been announced, but the present was already carrying weak structure from the trajectory that would later become undeniable. The person did not pull a fact out of nowhere. They were reading residues, however imperfectly.

Or take health. A body begins to drift before a diagnosis puts a clean label on the drift. The dramatic version says the person knew what was coming in some mysterious way. The stricter version says the body was already carrying local traces of a process in motion. Energy changed. recovery changed. appetite changed. tolerance changed. The person did not have full access. They had residual evidence. Their experience was inference from weak local structure, not a miracle.

The same is true in markets, in weather, in social life, in the feel of a room, in the first signs of a machine starting to fail. In each case the future does not need to broadcast itself backward. The present is already being shaped by what is underway. The key is that what reaches the observer is rarely complete. It comes as leftovers. Fragments. Hints. Disturbed patterns. Partial order.

That is why the word local matters so much in this definition. LERI is not about distant omniscience. It is about what is available here, now, through actual contact. Local air pressure. Local vibration. Local timing shifts. Local behavior change. Local record. Local trace. Local noise mixed with local signal. If the information is not locally available in some form, then LERI has nothing to work with. The claim does not get stronger when it leaves the local world behind. It dies.

This local emphasis also protects the idea from inflation. People often think a serious concept has to sound larger than life. In reality the strongest concepts usually become stronger by becoming smaller. They name exactly what is required, and they refuse everything else. LERI requires only that locally accessible reality may contain weak, structured residues of broader trajectories. That is enough. It is enough to explain why anticipation can sometimes feel real without turning it into prophecy. It is enough to explain why weak signals matter without pretending they are full explanations. It is enough to preserve the interesting part while discarding the mystical decoration.

There is another important limit built into the idea. Even when local residual traces exist, they are not automatically usable. A trace can be real and still too faint. A pattern can be present and still too noisy. A residue can survive and still remain below the threshold where a human mind, a sensor, or a model can reliably extract it. This matters because people often slide from there is information somewhere to therefore I should be able to use it. That does not follow. LERI only says that weak structured leftovers may exist and may sometimes support inference. It does not say those leftovers must always be detectable, legible, or operationally helpful.

In that sense LERI belongs naturally inside the larger logic of this book. Future Access Without Presence is already about the separation between what can be known and what can be moved. LERI extends that separation one step further. It says even knowledge itself comes in layers. There is full reality, then there is local access to full reality, then there is the degraded residue of that access after noise, delay, and filtering have done their work. By the time the observer receives the result, what remains may be enough for a weak inference while still being nowhere near enough for certainty, let alone control.

This is why the concept should feel disciplined rather than grand. It is not opening the door to limitless claims. It is describing a narrow regime in which weak prediction can arise from imperfect present traces. That is all. A person does not need to leave the present. The present is already richer than conscious language usually admits. Processes start before they announce themselves. Trajectories cast local shadows before they fully arrive. The world leaks clues about its own direction, not because it is trying to speak in riddles, but because causes and constraints often begin shaping the near field before they mature into outcomes.

So if we had to state the idea as plainly as possible, it would be this.

LERI means limited inference from locally available residual traces that still carry weak structure about a larger trajectory.

Limited, because the access is partial.

Inference, because the observer reasons from clues rather than possessing the hidden fact itself.

Locally available, because there must be a real present path.

Residual traces, because what survives is leftover structure, not the whole event.

Weak structure, because the signal is degraded and may be fragile, noisy, and hard to use.

A larger trajectory, because the clue points beyond itself without becoming identical to the thing it suggests.

That may sound less glamorous than the dramatic versions, but it is far more powerful. It tells us what kind of claim could actually survive contact with reality. It tells us why some anticipations feel meaningful without requiring magic. It tells us why the world can become legible before it becomes obvious. And it tells us, most of all, that the right way to think about weak access is not as a violation of ordinary limits, but as an unusually careful reading of what those limits still allow. Even if a trace is real, that does not mean it arrives in a form that can still be used. This is where many people lose the thread. They hear that the world may contain weak residual structure and they immediately imagine that whatever is there should remain clean, visible, and available to thought. But information does not travel like a royal guest carried untouched across a polished floor. It travels through contact, through friction, through interference, through distance, through imperfect media, through channels that alter it while they carry it. By the time something reaches an observer, it may still be present in one sense and already badly wounded in another.

This is the real heart of readout. Not whether structure exists in the large sense, but whether enough of that structure survives the journey into local contact. A thing can be real and yet become increasingly useless as it passes through the world. That is not exotic. It is one of the most ordinary facts in life. We live by the remains of signals, not by their original purity. We live on what survives.

Think of a photograph copied again and again. The first image may be sharp. The tenth still contains the same scene in some abstract sense, but edges blur, contrast fades, small details smear into one another, background noise rises, the meaningful features become harder to separate from the defects of the copying process itself. Nothing magical happened. The picture did not become false simply because it became degraded. The important point is subtler. The original structure remains partly there, but the readout becomes less reliable. At some point a real image turns into a doubtful one, then into a suggestive one, then into something that may still contain information but no longer enough to support confident use.

That is how many traces behave. They do not vanish in one dramatic instant. They decay. They become weaker, noisier, less distinct from the surrounding clutter. The signal may still exist, but the cost of extracting it rises while the confidence of extraction falls. This is why the phrase residual access must always be handled with care. What survives is not the thing in its own strength. What survives is what the channel failed to destroy.

A rumor chain offers another clean analogy. One person sees something clearly and tells the next person, who tells the next, who tells the next. Even if no one intends to lie, the message shifts. Tone gets added. Emphasis changes. Uncertainty hardens into detail. Detail dissolves into vagueness. Some parts are repeated because they are emotionally striking. Other parts are dropped because they are boring, awkward, or difficult to remember. After enough links the chain still points back to a real event, but the event is now wrapped in distortion. Information remains, yet readout quality has fallen. The message has not become pure fiction, but neither does it remain a faithful transport of the original scene.

This matters because many forms of anticipation are built from exactly that sort of degraded contact. The observer is not reading the source event directly. The observer is reading the aftereffects of the aftereffects, small marks left by processes already in motion, fragments of a pattern that has passed through many filters before it becomes experience. What arrives may still carry a directional hint. It may still bias expectation. It may still alter feeling. But that does not mean it arrives with the cleanliness people often imagine when they talk about knowing.

A weak microphone in a noisy room makes the same point in a more physical way. A speaker may be saying something perfectly intelligible, but if the room is full of interference, if the microphone is poor, if the wiring is damaged, if the output speaker crackles, then the listener receives a damaged version of the message. Some words come through. Others collapse into static. Maybe the general subject is preserved. Maybe the emotional tone survives. Maybe a few sharp phrases make it intact. But much is lost. The listener is now performing inference on damaged arrival. They are reconstructing from fragments. They are not hearing the clean source.

That is exactly what readout degradation means. It does not mean there is no information. It means information arrives mixed with everything that happened to it along the way. The world is not a perfect delivery service. A present accessible trace may still contain structure, but it contains it after passage through delay, masking, dilution, contamination, and partial loss. This is why one person may pick up something vague, another nothing at all, and another only a feeling without clear explanation. The channel is not handing over a polished report. It is handing over what survived.

A noisy screen behind a wall gives another version of the same lesson. Imagine a bright display in the next room, but you do not have direct line of sight. You only receive stray flicker through a crack, or reflected light on the floor, or faint color shifting along the edge of the door. Something is clearly there. The neighboring room is not empty. The changing pattern of light proves that. But your access is highly constrained. You cannot read the full image. You cannot cleanly recover the text on the screen. You may infer motion, intensity, rhythm, perhaps even something about the kind of material being shown. Yet your readout remains degraded because the path is degraded. The question is never only what exists on the other side. The question is what your local channel lets through.

This is why attenuation matters so much. Attenuation is simply the weakening of what travels. The farther a trace moves from its source, the more it tends to lose force. But distance is not only spatial. It can be temporal, behavioral, sensory, biological, environmental, or computational. A signal can weaken because too much time passed. It can weaken because too many layers intervened. It can weaken because other processes drowned it out. It can weaken because the medium that carried it was poor. In every case the result is similar. Less of the original structure survives in usable form.

Noise makes the problem even worse. Noise is not merely distraction. It is everything mixed into the channel that was not part of the specific structure you hoped to read. Real life is full of noise. Mood is noise. Memory bias is noise. Stress is noise. Random fluctuation is noise. Background motion is noise. Unrelated changes in the environment are noise. Habitual expectation is noise. Wishful thinking is noise. Fear is noise. By the time a local observer receives a trace, they are not separating signal from silence. They are separating signal from a crowded field of competing fragments.

This is why degradation is not only about loss. It is also about confusion. A weak trace can remain strong enough to affect the observer while being too weak to interpret correctly. That is a dangerous zone. It is where a person may feel something before they can say what it is. It is where anticipation may become possible without clear explanation. It is also where false positives grow. The signal is not absent, but the channel no longer preserves it cleanly enough to make readout easy. This can produce real sensitivity and real error at the same time.

Human experience is full of such middle zones. You walk into a place and feel the mood instantly, but you cannot yet explain which cues created the feeling. You sense that someone is about to leave long before they say the words, but you cannot isolate the exact chain of evidence. You feel that your body is not moving on the old terms, but the symptom pattern is still too muddy to name. In each case the mind is working on degraded readout. Something in the present carries structure. Enough survives to bend expectation. But the path is imperfect, and the result enters awareness as tension, drift, weight, asymmetry, or unease rather than as a clear articulated fact.

This is where the language of fidelity becomes important. Fidelity means how well the arriving trace preserves the structure that mattered in the source. A low fidelity trace may still point in the right direction while losing almost all of its detail. A high fidelity trace preserves enough of the pattern to support more confident inference. The whole problem of readout lives inside that difference.

Not all surviving signals are equal. Some are too degraded to do anything but create atmosphere. Others preserve enough shape to support action. Many hover in between.

That hovering middle is exactly why this chapter matters. Once you understand degradation, weak access stops sounding mystical and starts sounding fragile. It becomes easier to see why traces must be handled cautiously. The observer is not sitting before pure truth. The observer is sitting before a damaged arrival, trying to separate what remains from what the channel has added, erased, or warped. Sometimes the result is still useful. Sometimes it is suggestive but thin. Sometimes the residue is real and yet no longer operationally meaningful.

That is the sober point to keep in view. Information can survive while becoming too degraded to guide action well. Structure can remain present while becoming practically unreadable. A trace can be real and yet nearly useless. This is not a contradiction. It is the normal fate of signals in a finite world. Once that is understood, the rest of the argument becomes much cleaner. Weak access is never about untouched truth somehow appearing in the mind. It is about whatever faint order remains after the world has done its best to blur it.

Once the idea has been narrowed this far, a fair question appears almost immediately. Fine, suppose locally available residual traces can in principle carry weak structure. Suppose degraded readout is possible. Suppose some anticipations are really inferences from damaged present clues rather than dramatic acts of foresight. How would you tell whether any of that is real, instead of just being another elegant story told after the fact.

That is where the experimental side becomes decisive. A concept like this cannot be left at the level of atmosphere. It has to be pushed until it either behaves like a real readout theory or falls apart under pressure. The reason the E LERI suite matters is that it tries to do exactly that. It does not ask whether the idea sounds deep. It asks whether weak access behaves the way a genuine residual readout process should behave. That is a much harder standard, and that is why it is useful.

The logic of the suite is simple in spirit, even if the implementation can become technical. If residual access is real, then it should show three things. First, there should be a coherent baseline picture that shows what weak readout is even supposed to mean in operational terms. Second, detectability should survive proper controls rather than appearing only in uncontrolled measurements. Third, the effect should weaken or disappear when the readout path itself is made worse. Those three steps sound modest, but together they draw a sharp line between disciplined inference and rhetorical fog.

The first step is the sanity check. Before anyone talks about residual access in a serious way, the underlying logic has to be made visible in a stripped down setting. That means showing, in a controlled and transparent form, how an observer can receive usable information through a weak and limited channel without confusing that information with full access. This matters because without a baseline model, everything becomes too verbal. People start arguing about the idea in abstract terms, and abstraction is where confusion multiplies. A sanity check gives the reader something firmer. It says, here is a toy case, here is the channel, here is the degradation, here is what survives, here is what does not.

That kind of first step does not prove a rich real world claim by itself, and it is not supposed to. Its purpose is narrower. It establishes that the concept is internally coherent. It shows that the idea of limited present readout is not nonsense on its face. It separates the small legitimate claim from the larger illegitimate ones. In effect it says, weak access can be formulated without magic, without direct future contact, and without asking the observer to become larger than the local signal path.

But coherence is only the entrance fee. Plenty of bad ideas can be stated coherently. The second step is much tougher. Once you claim there is a detectable trace, you have to ask whether you are seeing a real effect or merely flattering noise. This is where null calibration and persistence gating matter so much. A weak signal is exactly the sort of thing that can be faked by loose methods. If the effect only appears when the measurement is allowed to drift with hidden structure, selective averaging, accidental alignment, or wishful interpretation, then it tells us very little. A disciplined suite has to ask what the background looks like when the relevant alignment is broken, and whether the apparent access remains meaningfully above that background.

That is why null controls are not decorative. They are the difference between an interesting result and a mood. In plain language, a null control asks what happens when the alignment that would carry real information is intentionally disrupted while as much irrelevant structure as possible is preserved. If the apparent effect survives only because the whole system contains broad trends or built in regularities that would produce a false appearance of signal anyway, then the null will expose that. If, on the other hand, the effect collapses under the null while remaining present in the properly aligned measurement, then the claim becomes much more credible. It starts to look less like storytelling and more like actual readout.

Persistence gating adds another layer of seriousness. A true weak effect should not be defined by one lucky blip. It should show some stability across neighboring settings, repeated trials, or closely related conditions. This matters because noise loves isolated peaks. Randomness produces dramatic points all the time. Human beings are then tempted to point at the peak and say there, that is the thing. But a real readout process should have more shape than that. It should not require the observer to worship a single accident. Requiring persistence is a way of forcing the signal to prove that it belongs to a pattern rather than a moment of excitement.

Together, null calibration and persistence gating do something very important. They turn the question from can I tell a compelling story about this weak effect, into does the effect survive contact with methods designed to embarrass it. That is the right standard. A concept like LERI should not be protected from humiliation. It should be exposed to it. If it is real, even weakly real, it should survive that kind of pressure better than mere coincidence, drift, or memory bias.

Then comes the strongest test of all, the shielding sweep. In a way this is the heart of the entire package, because it asks the one question weak access theories cannot avoid. If the observer depends on residual readout, what happens when the readout channel itself is weakened.

This is where the idea becomes properly mechanical. Not mechanical in the sense of dull, but in the sense that it finally acquires a hard consequence. A mystical claim can survive any condition because it never truly committed itself to a channel in the first place. It can always retreat into vagueness. A real readout theory does not have that luxury. If the effect depends on weak local structure reaching the observer, then reducing the gain of that path, adding shielding, increasing attenuation, or otherwise worsening the channel should reduce detectability. At some point the effect should collapse.

That is a beautiful constraint because it forces honesty. It says, do not merely tell me that the signal exists. Show me that its operational presence tracks the quality of the path. Show me that when the observer is separated more strongly from the trace, the observer loses usable access. Show me that the result is not free floating. Show me that it lives and dies with the channel that supposedly carries it.

This is exactly the sort of behavior one would expect from residual readout if the concept means anything at all. A dim light behind a wall becomes harder to detect when the wall gets thicker. A weak smell becomes harder to detect when the air path is blocked. A faint vibration becomes harder to feel when the contact surface is damped. No one would find that surprising. Yet this obviousness is precisely what gives the shielding result its force. It moves the whole discussion back into the ordinary logic of signal path and readout. The observer gets less because less arrives. That is the kind of sentence a serious theory should want to be able to survive.

The strength of the E LERI suite, then, is not that it tries to prove something spectacular. Its strength is almost the opposite. It tries to show that weak access can be treated as a modest measurable phenomenon with expected failure modes. A weak trace should be detectable only under some conditions. It should sit above calibrated background rather than vague enthusiasm. It should require more than one dramatic point. And it should weaken when the readout path itself is weakened. That combination does not make the claim large. It makes it responsible.

This is also why the suite fits so naturally into the broader architecture of the book. The whole project of FAWP depends on separating categories that people habitually mix together. Prediction is not steering. Presence is not influence. Existence is not access. And here we add another layer. Residual structure is not the same thing as usable readout. There may be information in the world that does not reach the observer in a form they can actually use. There may even be information that reaches them weakly enough to bias anticipation while remaining far too degraded for certainty. The suite is built to study exactly that narrow territory.

Seen that way, the experiments do not function as decoration around a philosophical idea. They function as discipline. They say, if you want to talk about weak access, then you must also talk about thresholds, controls, persistence, attenuation, and collapse. You must talk about when the effect appears, when it vanishes, and why. You must accept the possibility that some traces are too weak to matter, some apparent signals are artifacts, and some beautiful stories are only stories. That is not a loss. It is what saves the concept from becoming useless.

So the real achievement of the E LERI suite is not that it gives the reader a thrilling new power fantasy. It does something more valuable. It takes a territory that usually invites inflation and forces it to obey the logic of damaged channels, calibrated detection, and degraded access. It teaches the reader to ask the right question. Not did something uncanny happen. But what path existed, what survived the path, and what happened to the effect when the path itself was made worse. Once you start asking that question, most of the cheap mystery drains away, and what remains is smaller, stricter, and far more interesting.

The most convincing ideas are usually the ones that become weaker in exactly the way they are supposed to become weaker. That may sound strange at first, because people often think strength in an idea means resilience under every condition. They imagine that a powerful claim should survive any pressure, any change of setting, any attempt to block it. But that is not how real signal based explanations behave. Real signal based explanations are strong precisely because they are vulnerable in lawful ways. If you damage the path, the effect should weaken. If you block enough of the channel, the effect should collapse. If you increase interference, readout should become less reliable. That is not a problem for the theory. That is the theory becoming more believable.

This is why shielding matters so much.

If weak access depends on local residual traces, then the quality of the local path cannot be a side detail. It has to sit near the center of the whole picture. The observer is not drawing information out of emptiness. The observer is not bypassing the world. The observer is not receiving a perfect delivery from nowhere. The observer is working with whatever the channel still allows to survive. That means the channel itself must matter. Not poetically, but operationally. Not as a metaphor, but as a condition.

The idea becomes very simple once stated clearly. If a person can only detect a faint signal because some weak structured trace is still reaching them, then reducing the strength of that arrival should reduce what they can detect. If less of the relevant structure gets through, less should remain available for inference. At some point the surviving pattern should become too weak, too noisy, or too damaged to matter. That is what shielding tests.

Think about a lamp behind a curtain. If the curtain is thin, the light still leaks through. You may not see the bulb itself, but you can tell something bright is there. The room is altered by it. The wall glows faintly, the edge of the fabric lights up, the surrounding darkness is no longer uniform. Now add another curtain. Then another. Then heavier material. What should happen. The answer is obvious. The local evidence should diminish. The glow should weaken. The shape should blur. The confidence of detection should fall. Eventually the lamp may still exist exactly as before, but for the observer on this side of the barrier it becomes operationally absent. Not nonexistent, not erased from the larger reality, but inaccessible in any usable way.

That is the whole point in miniature. Shielding separates existence from access with brutal clarity. It shows that something can remain real while becoming unreadable. It shows that the difference between a detectable trace and an undetectable one may have less to do with the deeper truth of the source than with the quality of the path between source and observer. It returns the whole discussion to the ordinary logic of signal and barrier.

This is also why shielding is stronger than anecdote. Anecdote can tell you that something seemed to happen. Shielding asks whether the supposed effect behaves like something that depends on a real local path. That is a much tougher demand. A mystical story can survive any environment because it does not commit itself to a mechanism. It can always retreat into language about hidden forces, deeper levels, or inaccessible causes. A readout theory does not get that luxury. The moment it says local residual traces matter, it has accepted a burden. It now has to weaken when the local path weakens. If it does not, then the claim begins to look hollow.

This is why a shielding result is such a clean stress test. It forces the idea to stop floating. It ties the claim to an observable dependence. More barrier, less readout. More attenuation, less detectability. More channel damage, less usable structure. That is exactly what should happen if the observer is relying on degraded local evidence rather than direct unrestricted access.

The beauty of this is that it makes the theory smaller and therefore stronger. A smaller claim can be tested more honestly. A larger claim hides in vagueness. People often think mystery becomes deeper when its boundaries disappear. In science and in thought, the opposite is usually true. Mystery becomes more respectable when it accepts a border. Shielding provides that border. It says there is no free access here. There is only what survives the path.

Once you see that, many familiar confusions start to dissolve. People often speak as though a real signal should either be fully available or not available at all. But shielding reminds us that access is graded. It can thin out. It can degrade. It can remain barely present. It can pass below practical detectability without any dramatic break in the underlying reality. This matters because the world is full of transitions like that. Sound fades with distance. Heat disperses. Light weakens. Smell diffuses. Pressure dissipates. A signal based theory of weak access should look like those things. It should not behave like a supernatural switch.

That graded quality is important because it explains why some effects may feel frustratingly unstable. A trace near the threshold is not fully secure. Small changes in the path may matter a great deal. A little more interference, a little more masking, a little more attenuation, and the observer loses usable contact. That does not mean there was never anything there. It means the connection was fragile, and fragile things vanish quickly when the conditions worsen.

This is one reason shielding tells us something deeper than simple blockage. It tells us about dependence. If detectability falls in step with reduced channel quality, then the channel was doing real work. The effect was not floating free of the environment. It was riding on it. That is exactly what a serious theory should want to see. Not invincibility, but lawful dependence. Not theatrical persistence, but structured collapse.

The phrase structured collapse is worth holding onto. A bad idea often fails chaotically. It gives one strange result here, another somewhere else, and no one can explain why it appears or vanishes. A better idea fails in recognizable ways. It loses strength where the channel is damaged. It disappears where the threshold is crossed. It declines with worsening conditions. Its weakness has a pattern. That pattern is not an embarrassment. It is evidence that the explanation is tied to the architecture of the situation.

In that sense shielding is not merely a technical detail in the readout story. It is the point where the story becomes honest. It forces the concept to admit that access has a cost, a path, and a limit. It strips away the temptation to say that because some larger structure exists, the observer should somehow remain connected to it no matter what stands in the way. No. If the path narrows, access narrows. If the path breaks, access breaks. The world remains what it is, but the observer receives less of it.

That lesson matters far beyond this one topic. It echoes everything this book has been arguing from the beginning. Prediction is not control. Presence is not influence. Existence is not access. And now we can add one more line. Residual structure is not usable readout unless enough of it actually survives the journey. Shielding makes that truth impossible to ignore.

It also protects the reader from a dangerous emotional mistake. When people hear about weak access, they often imagine that if they cannot recover a signal cleanly enough, then the signal must never have existed. But that is not how finite systems work. Something can be there and still be beyond your effective reach because the path has become too poor. The trace can be real and still not survive in a form you can trust. The failure belongs to the channel, not necessarily to the source. Shielding teaches that distinction with unusual force.

And perhaps most importantly, it gives the whole discussion a kind of dignity. It stops weak access from becoming a fantasy of hidden power. Instead it becomes what it really is, if it is anything at all. A narrow, fragile, conditional form of contact, one that lives only so long as a real path keeps some faint order alive. Thicken the barrier enough, and the effect dies. Not because the universe changed its mind, but because the observer no longer has enough left to read.

That is exactly the sort of limit a serious idea should accept. It does not make the claim smaller in the cheap sense. It makes it smaller in the disciplined sense. It replaces wonder without boundaries with a weaker but more honest kind of wonder, the wonder that comes from seeing how much the world can still reveal through a damaged path, and how quickly that revelation fails once the path is made just a little worse.

At this point the most important thing is to say, with complete plainness, what this idea does not claim. That matters because almost every subject in this territory becomes corrupted at the moment people start rewarding it for sounding bigger than it is. The whole value of a careful concept is that it keeps refusing those rewards. It gives up glamour in exchange for clarity. If it cannot do that, then it is not a useful concept at all.

So let us begin with the clean refusal. This idea does not claim prophecy. It does not claim that a person stands outside ordinary limits and receives tomorrow as a finished object. It does not claim that the mind can simply reach across time and collect distant facts at will. It does not claim that every strong feeling is meaningful. It does not claim that coincidence deserves worship. It does not claim that anxiety is insight, or that vividness is evidence, or that emotional intensity should be mistaken for access. It does not claim hidden powers in the dramatic sense. It does not claim a secret exemption from the ordinary damage done by noise, distance, delay, filtering, and weak channels.

In one sense this sounds like reduction, and it is reduction. But it is the kind of reduction that saves an idea instead of killing it. An inflated claim becomes impossible to test because it no longer knows what it needs in order to be true. A smaller claim knows exactly what would support it and exactly what would expose its weakness. That is the position worth keeping.

The narrower claim is enough. It says that locally available traces may carry weak structure. It says that such structure may sometimes support limited inference. It says that what reaches the observer is partial, degraded, and path dependent. It says that readout can fail. It says that apparent access should weaken when the channel weakens. It says that what matters is not whether reality contains hidden order in the grandest sense, but whether enough of that order survives into local contact to make any practical difference at all.

Once the claim is stated that way, something important happens. The subject stops floating free of the rest of the book and snaps directly into place inside it. This is because the broader argument has never been about unlimited access. It has never been about godlike reach. It has never been about certainty. From the beginning, the argument has been about separation. Prediction can survive where control has weakened. Legibility can survive where leverage has collapsed. The future can become inferable before it becomes touchable. LERI belongs here because it sharpens that same family of distinctions instead of softening them.

In fact it adds one more layer of discipline to the whole picture. It tells us that even prediction itself has to be treated carefully. Not every form of apparent knowing is the same. There is a difference between clean present evidence, weak residual evidence, post hoc storytelling, wishful anticipation, and noise that only looks meaningful after memory edits the scene. By making that difference explicit, LERI does not open the floodgates to bigger claims. It closes them. It says prediction must also answer to channels, thresholds, and failure modes. That is exactly what the book needs.

This is why the concept strengthens Future Access Without Presence instead of weakening it. A weak version of the book would drift into vague statements about people somehow sensing things. A stronger version has to say much more. It has to say what kind of signal might be involved, what kind of path would have to exist, what kind of degradation would be expected, and why the effect should disappear under the wrong conditions. That makes the overall argument harder to misunderstand. It keeps the line clean between a future that is partly inferable and a future that is somehow being directly delivered in finished form. The first claim is difficult but manageable. The second is mostly theater.

There is a deeper reason this matters. The whole emotional force of FAWP comes from a human experience that does not need magic in order to be profound. People already know what it feels like to live inside a world where the future begins to announce itself before they can meaningfully redirect it. They know what it feels like when the body whispers before it breaks, when love cools before it ends, when work hollows before it collapses, when markets become unstable before they correct, when public life grows brittle before institutions respond. None of that requires direct future access. It requires only that trajectories leave marks before they become official events. That is enough. In fact it is more than enough. It is the daily substance of the argument.

LERI helps protect that truth from two bad habits. The first bad habit is inflation. That is the urge to take every weak signal and turn it into cosmic drama. The second bad habit is dismissal. That is the urge to say that unless access is perfect, there was no access at all. Both habits are forms of laziness. Inflation makes the claim too large to survive honest pressure. Dismissal makes the world flatter than experience justifies. The disciplined middle path is better. It says the world can be partially readable without becoming transparent. It says weak access can exist without becoming direct presence. It says a person may receive enough residual structure to lean toward what is coming, while still receiving far too little to deserve certainty.

That is the exact kind of nuance this book needs. Future Access Without Presence is not about omniscience without control. It is about partial legibility without proportionate leverage. The whole framework becomes stronger the moment the reader understands that partial legibility itself can come in damaged layers. Full reality is one thing. Local access to full reality is another. Degraded residual traces are another still. By the time the observer stands inside that final layer, what remains may be enough to suggest, enough to bias expectation, enough to create unease, enough to support weak prediction, and still nowhere near enough to count as possession of the truth. That is not a flaw in the idea. It is the human condition stated with more accuracy.

There is also an ethical gain here. A concept like this, if left vague, invites vanity. People love the idea that they are secretly ahead of the world. They love the image of hidden gifts, private certainty, and subtle superiority over the ordinary limits of evidence. LERI cuts against that vanity. It reminds the reader that whatever weak access exists, it is fragile, conditional, local, and often badly damaged. It reminds them that the channel can fail, that the trace can be too faint, that readout can be wrong, that memory can flatter itself, and that the observer remains finite. That humility is not a side benefit. It is one of the reasons the concept is worth keeping at all.

At the same time, humility does not mean emptiness. The smaller claim still matters. It matters because it helps explain why some forms of anticipation feel real without requiring impossible explanations. It matters because it keeps the present richer than a flat instant. It matters because it gives language to the way processes often begin before they declare themselves. It matters because it helps us see why a future can start influencing perception before it enters life as an undeniable event. Most of all, it matters because it fits the book's central insight. The world often becomes readable before it becomes reachable. LERI simply tells us that readability itself must obey the harsh limits of readout.

So the final lesson is not that the world is magical. The final lesson is stricter, and in its own way more beautiful. Reality may contain more structure than the observer can cleanly use. Some of that structure may survive into local traces. Some of those traces may support weak inference. But none of this abolishes limits. Access remains filtered. Readout remains damaged. Noise remains real. Influence remains separate. The observer is still a finite creature standing inside a partial channel, trying to recover a little order from what survives.

That is enough. More than enough. It lets the book end this difficult section on firm ground. The point was never to expand the claim until it sounded impressive. The point was to reduce it until it became honest. Once that reduction is complete, the larger thesis stands more clearly than before. The future can be partly inferable without being directly present. It can be partly legible without becoming controllable. It can leak into the present as trace, tendency, residue, and weak pattern, while still remaining beyond easy reach.

That is why the real question was never whether reality contains hidden order. The real question was whether enough of that order reaches a local observer, in time, through a real path, with enough surviving fidelity to matter. Everything in this section leads back to that one demand. Not bigger claims, but cleaner ones. Not fantasy, but damaged contact. Not revelation, but constrained readout. And once that is understood, the whole architecture of Future Access Without Presence becomes tighter, calmer, and harder to misunderstand.

Chapter 7

Most people imagine that the hardest part of life is not knowing. They imagine darkness, confusion, blind risk, total uncertainty. They think the real danger is walking into a bad future without any warning at all. That does happen. Sometimes life strikes cleanly and without notice. Sometimes the floor really does give way. But much of modern life feels bad for the opposite reason. The future does not stay hidden. It leaks. It signals. It arrives in fragments. It leaves traces before it fully appears. You can often feel where things are heading before you can do very much to change the direction.

That is one reason modern stress feels so strange. It is not only panic. It is not only ignorance. It is often prolonged exposure to visible trajectories. You are not fully in the dark, but you are not fully in control either. You can sense the shape of the thing before the thing becomes official. The pattern becomes legible before it becomes easy to interrupt. The signal arrives before the handle does. That is the atmosphere in which a lot of ordinary people now live.

A person notices that their body has changed, but the change is still small enough to dismiss in public. A person feels their relationship cooling, but nothing dramatic has happened yet. A person can tell their job has started to drift toward irrelevance, but no formal announcement has arrived. A person watches prices rise, debts tighten, and margin disappear, but the numbers still do not look catastrophic enough to force a reset. A person sees a whole culture becoming more brittle, more tired, more performative, more fragmented, but there is no obvious single lever to pull. In each case the same condition appears. The path ahead is no longer invisible, but the practical channel for changing it is weak, delayed, expensive, or scattered across too many moving parts.

This is why so many people feel both informed and powerless at the same time. They are not stupid. They are not always passive.

They are not always lacking effort. Very often they are trapped in an environment where awareness has expanded faster than influence. They can see more than earlier generations could see. They can track more variables, hear more warnings, compare more trends, absorb more commentary, detect more weak signals. But greater visibility does not automatically create greater leverage. In many cases it does the opposite. It creates a world in which you can witness your own narrowing options in high resolution.

That tension runs through almost every serious domain of modern life. The body gives hints before it gives verdicts. A bond begins to fail before it breaks. A workplace decays before it collapses. A market becomes unstable before it corrects. A platform learns your behavior faster than you learn how it shapes you. A city becomes less affordable before it becomes impossible. A society becomes more exhausted before it openly admits exhaustion. The early stage is rarely a clean emergency. It is usually a long middle. That middle is where many people now spend their lives. Not in perfect safety, not in total crisis, but in an extended experience of watching the future gather force before they can bring enough force of their own to meet it.

This matters because people often misread that condition as weakness. They think that if they can see the problem, they should be able to act on it. If they understand the direction, they should be able to steer. If they feel the wave early, they should be able to move out of its path. But that assumes that prediction and control are the same thing, and much of this book has been devoted to showing that they are not. The world can become more readable while your local action becomes less effective. The pattern can strengthen while your leverage weakens. The signal can sharpen while the cost of changing course rises. Once you see that clearly, a lot of familiar human frustration stops looking like a character defect and starts looking like a structural feature of life under delay, noise, scale, and distributed consequence.

You can see this even in the simplest daily routines. Think about how often people say things like, I knew this was coming, I could feel it months ago, I saw the signs, I had a bad feeling, nothing was official yet, but something was off. Those phrases are everywhere. They appear after breakups, layoffs, diagnoses, financial strain, burnout, family breakdown, public decline, and personal drift. People say them because they are trying to describe a state that is both clear and hard to justify. They are describing the moment when reality is not yet undeniable, but has already stopped feeling neutral. They are describing the uneasy territory where visibility begins before authority, proof, or social permission catches up.

That territory is one of the defining emotional climates of the present. Earlier eras were full of uncertainty too, of course. But modern life has amplified a very particular kind of uncertainty, uncertainty in full view. You can now see trends, compare lives, monitor systems, track statistics, follow markets, read cultural moods, watch institutions wobble, and notice the first symptoms of strain almost in real time. Yet the same world that increases visibility often stretches action across layers of friction. Decisions are delayed. Systems are large. Consequences are distributed. Responsibility is diffuse. Social response is slow. Personal attention is fragmented. The result is not simply more knowledge. It is more knowledge under weaker local control.

That is a recipe for a very specific kind of human fatigue. It is the fatigue of prolonged recognition without clean intervention. You know enough to worry, but not enough to command. You know enough to prepare, but not enough to relax. You know enough to sense drift, but not enough to stop the machine on your own. This is why so many modern people feel permanently braced. They are not just reacting to isolated shocks. They are living inside extended preludes. They are carrying the weight of futures that have already started to announce themselves, but have not yet arrived in forms that make decisive action feel simple.

This condition shows up in private life first, because private life is where the human nervous system feels these pressures most directly. The body begins to whisper before it shouts. Love begins to thin before it disappears. Money begins to tighten before it breaks routine. Work begins to lose meaning before it loses structure. Habits begin to harden before they become visible to others as a problem. But the same condition also appears at larger scales. Entire populations can sense that something is changing long before any single institution responds with clarity. People can feel that trust is lower, energy is flatter, conversation is harsher, attention is thinner, and long term confidence is weaker. These are not always dramatic signals, but they accumulate. A future can begin emotionally before it becomes formally real.

That is why Future Access Without Presence is not some niche curiosity. It is not a rare edge case that appears only in abstract models or specialized experiments. It is one of the most common textures of lived experience in a world where signal travels fast and influence often travels slowly. It appears wherever pattern recognition survives longer than practical leverage. It appears wherever the future begins to reveal itself before the reachable window stays wide enough to feel comfortable. It appears wherever the map gets better while the steering wheel gets worse.

The deepest point is not that life has become hopeless. The deeper point is that modern life has become legible in a new way. We are surrounded by indicators, traces, previews, weak warnings, and visible trajectories. We can often tell where the road bends before we get there. But seeing the bend is not the same thing as taking it well. In many parts of life, the future now becomes available to perception earlier than it becomes available to control. That difference is not a side issue. It is one of the central realities of contemporary existence.

Once you understand that, the ordinary world starts to look different. Health stops being just a matter of sudden bad news and starts looking like a sequence of signals, some early, some late, some actionable, some not. Relationships stop looking like clean events and start looking like long changes in warmth, trust, attention, and timing. Careers stop looking like stable ladders and start looking like moving systems full of lag, drift, and delayed recognition. Markets stop looking like simple tests of intelligence and start looking like brutal arenas where being right and being early are not enough unless execution remains alive. Technology stops looking like neutral convenience and starts looking like a structure that can increase prediction while thinning out personal control.

This is where we begin. Not in theory now, but in life itself. The question is no longer whether Future Access Without Presence exists. The question is where it appears, how it feels, and why so much of ordinary modern experience seems built from exactly this split between visibility and reach.

Health is one of the clearest places where this pattern shows up, because the body almost never tells its story all at once. It hints, it nudges, it drags, it interrupts, it changes the cost of ordinary things. It does not always arrive with a clean declaration. More often it begins with a subtle loss of ease. Sleep stops feeling restorative. Recovery slows down. Patience thins. Appetite shifts. The stairs feel slightly different. The afternoon becomes heavier. A drink hits harder. A missed night of rest costs more. The same life continues on paper, but the body has already started revising the terms.

People know this feeling very well. They say something is off. They say they do not feel like themselves. They say they are tired in a different way. They say it is probably stress, probably age, probably a phase, probably nothing serious. Sometimes they are right. Not every weak signal is a catastrophe. Not every drag is destiny. But the body is one of the great theaters of Future Access Without Presence because it often becomes legible before it becomes tractable in an easy, comfortable way. You can feel the direction before you know the name. You can notice the pattern before you have enough proof to make action feel socially justified.

That gap matters enormously. Health is not just a matter of what is true in the body. It is also a matter of when the truth becomes actionable. The earlier phase is often psychologically the hardest. The signals are real enough to disturb you, but not clean enough to shut down all doubt. You may feel foolish for taking them seriously. You may worry about overreacting. You may compare yourself to people who seem to push through similar symptoms without complaint. You may tell yourself that grown adults do not run to the doctor every time something feels strange. So you wait. Not because you are irrational, but because the body often presents its first warnings in forms that demand interpretation under uncertainty.

This is one reason health regret has such a particular sting. People do not only regret what happened. They regret the long stage in which they half knew. They look back and say, I felt something months ago. I knew my energy was not normal. I could tell my recovery was different. I had this feeling. That backward glance can become brutally self accusing. It creates the impression that the person had the truth in hand and simply failed to use it. But that is rarely the lived reality. More often they had partial access. The body was giving them information, but not yet in a form that made the path forward emotionally easy, cheap, or obvious.

This is exactly where the distinction between prediction and control becomes useful. A body can provide predictive information before it provides an easy control handle. You may be able to sense that a trajectory is wrong before you know which variable to change, which test to run, which habit matters most, or which threshold has been crossed. In some cases action is available and people delay it. In other cases action exists only in rough form. Sleep more, drink less, get checked, stop ignoring pain, stop normalizing exhaustion, take the symptom seriously enough to expose it to reality. Those are real moves, but they do not feel as satisfying as certainty. They feel provisional. They feel awkward. They feel like acting before the case is fully built. And that is exactly why so many people postpone them.

The body is also full of lag. This makes the problem worse. Many causes do not produce consequences immediately, and many consequences do not reveal causes cleanly. A person can live in a way that is slowly eroding them while daily life still appears functional. They can borrow energy from tomorrow and call it discipline. They can numb stress and call it coping. They can eat in ways that flatten attention, disrupt sleep, and distort mood without ever linking those changes together as one picture. They can live in chronic low grade strain for years, because the body is adaptive enough to keep the show running while quietly increasing the price of everything. By the time the person gets a result dramatic enough to force action, the story has already been unfolding for a very long time.

This is why modern health life can feel so strange. We have more data than ever, more tests, more trackers, more articles, more health content, more quantified indicators, more language for risk. In theory that should increase agency. Sometimes it does. Earlier screening saves lives. Wearables can expose patterns that would otherwise be missed. Better information can shorten the loop between feeling and action. But information alone does not solve the problem. In some cases it simply moves the ambiguity to a new level. A person watches their sleep score fall, their heart rate change, their stress measure spike, and now they have more visibility without necessarily having more leverage. They know more, but they are not always clearer on what matters, what is noise, what deserves action, or how urgently to intervene.

This creates a new form of health anxiety, one that is not built on pure ignorance, but on too much weak signal without stable interpretation. The person is not blind. They are half illuminated. They are surrounded by indicators, but still unsure which ones are driving outcomes and which ones are merely flickering. This can make the body feel even harder to manage, because awareness expands faster than clean action. You can monitor your decline in greater detail without yet knowing how to change the slope. That is a health version of Future Access Without Presence.

The medical world itself also reflects this split. A physician may recognize where a disease process is likely heading while having limited ability to reverse it at that stage. A patient may receive a warning that is real but probabilistic, serious but not immediate, actionable but not dramatic. Lifestyle advice often suffers from this exact problem. Move more, sleep better, reduce stress, change the diet, stop normalizing chronic overload. Everyone knows these sentences. The problem is not that the information is secret. The problem is that the body often turns gradual damage into delayed consequence. That delay weakens the emotional force of early action. It encourages the fantasy that the old pattern can continue a little longer without cost. Then one day the cost becomes undeniable, and now everyone speaks as if the issue has suddenly appeared.

But what makes health such a powerful example is that it shows how painful the timing problem really is. If a person waits until the body makes its case in perfect language, they often wait too long for the easiest kind of intervention. Earlier signals may have been softer, but they were also closer to the zone where change was cheaper. This does not mean all illness is preventable, or that every diagnosis is the result of missed signals. That would be cruel and false. Bodies are finite, luck is real, genetics are real, accidents are real, and biology can move against a person even when they do many things right. The point is narrower and more useful. In the parts of health that do depend on feedback, timing, and behavior, the body often gives you legibility before it gives you comfort, and waiting for comfort can cost you leverage.

That fact should change how we think about self trust. Self trust in health is not mystical intuition. It is not the fantasy that every feeling is a revelation. It is the disciplined refusal to downgrade repeated friction just because it has not yet become theatrical. It is the willingness to notice patterns before they are fully named. It is the courage to expose vague trouble to concrete reality. Get the test. Ask the question. Change the routine. Stop treating constant drag as the price of adulthood. Stop assuming that functional means healthy. A person can still be working, commuting, answering messages, paying bills, smiling in public, and living inside a body that has already begun to warn them.

This is why health is such a central example of the larger idea. The body becomes visible before it becomes obvious. Risk becomes knowable before it becomes dramatic. The path begins before the event. And because the body is intimate, because it is the place where all abstraction becomes lived cost, people feel this split with unusual force. They feel what it means to recognize a direction before they can easily interrupt it. They feel what it means for clarity to arrive in a form too weak to comfort and too strong to ignore. They feel, in the most personal way possible, what it means for the future to start leaking into the present before the handle arrives in the hand.

Work is one of the most common places where people live inside Future Access Without Presence, because jobs rarely disappear at the exact moment people begin to suspect danger. Most roles do not go from stable to gone in a single clean jump. They thin out. They lose gravity. They begin to matter less in ways that are visible long before they are official. The future of a person inside a company or institution often starts changing before the language around that change is allowed to become direct.

People feel this long before they can always explain it. The work that once carried energy starts arriving with less force. Promises become softer. Decisions begin happening farther away. A person notices that they are no longer at the center of what once required them. The role still exists, the pay still lands, the title still holds, the routine still looks recognizably normal from the outside. But something important has shifted in the pattern. The person is still standing inside the same structure, yet their actual contact with its future has weakened.

That is one reason job anxiety has such a peculiar tone. It is often not the fear of sudden catastrophe. It is the fear of prolonged drift under partial visibility. A person senses that their place is becoming less secure, less necessary, less connected to the real direction of things, but the signal is not yet clean enough to satisfy every inner skeptic. They tell themselves they may be imagining it. They tell themselves the slowdown is temporary, the reorganization is routine, the silence means nothing, the reduced priority is only a phase. Sometimes they are right. Large structures produce noise. Not every pause is a verdict. But work life is full of situations where the future of a role becomes legible before the person occupying it can still shape that future with confidence.

This is part of what makes careers so emotionally difficult to manage. Work is not merely a stream of tasks. It is also identity, stability, momentum, dignity, social standing, future planning, and daily structure. That means the signs of drift are often interpreted through hope as much as through evidence. A person may feel the role changing and still resist updating their model, because the cost of updating is too large. To admit that a job is weakening is to admit that a whole set of assumptions may also be weakening, income, timing, options, self story, and the ordinary picture of what the next year was supposed to look like. So people wait for stronger evidence. They want the case to become undeniable. They want reality to speak in a form that removes self doubt. But by the time work speaks that clearly, the easiest moves may already be gone.

This is where the split between prediction and control becomes painfully practical. A person can often forecast decline before they can cleanly steer around it. They can feel that a company is tightening, that a project no longer matters, that their skills are becoming less central, that the structure is slowly withdrawing real commitment, all before there is a simple lever to pull. The signals are there, but the action channel is awkward. Update your skills now, even though nothing official has happened. Explore exits now, even though leaving early may look paranoid. Reduce financial dependency now, even though the pay is still coming. Reposition now, while repositioning still feels emotionally unjustified. Those are real moves, but they are expensive precisely because they ask for action before certainty.

This is not only a personal problem. It is built into how large organizations behave. Big structures tend to absorb trouble slowly and reveal it unevenly. Problems are diluted across distance, timing, incentives, and language. A person near the edge can often feel that something is off long before anyone at scale names it plainly. Morale shifts first. Energy shifts first. Investment shifts first. Attention shifts first. The institution may already be turning toward a different future while still allowing the old future to remain verbally intact. That creates a painful environment for local agency. The people inside the structure are not fully in the dark, but neither are they standing in a place where their local effort can reliably alter the broader direction.

This is why so many workers end up living in a kind of suspended recognition. They know enough to worry, but not enough to act without inner conflict. They sense that the machine is changing, but the machine is large enough that their individual hand cannot easily test what is still possible. They may try to work harder, produce more, stay visible, remain positive, and anticipate needs. Sometimes that helps. Sometimes early initiative really does matter. But in many cases effort becomes uncoupled from outcome long before the person realizes it. They are still investing energy into a future that has already begun to close above them. They are still behaving as though performance alone governs the path, even though the path has become shaped by wider forces, cost cutting, strategic change, automation, market pressure, internal hierarchy, or a simple shift in where value is now being recognized.

That mismatch is one of the hidden brutalities of work life. People are often taught that if they are competent, committed, and adaptive, then stability will follow. There is some truth in that. Skill matters, reputation matters, timing matters, and people do create opportunities through excellence. But there is another truth that is harder to digest. A role can become less viable even while the person performing it remains strong. A person can be increasingly right for yesterday and increasingly exposed to tomorrow. They can feel the change before they can stop it. That is a career form of Future Access Without Presence.

Technology has intensified this problem rather than solving it. Many workers now live in environments where visibility has increased but leverage has not. They can see metrics, dashboards, performance data, industry trends, hiring changes, market commentary, and signals of internal drift almost in real time. In theory that should make adaptation easier. Sometimes it does. But just as often it means a person can watch the meaning of their work change faster than they can reposition their life around that change. They become highly informed about a process they are still too entangled to alter quickly. Awareness rises, but the friction of response remains large.

This is especially clear when skills begin to lose their old advantage. A person can often sense that the market value of what they know is changing before any single event forces the issue. The work that once felt secure now feels crowded, automated, commoditized, or quietly devalued. The person does not need a final verdict to notice this. They can feel it in rates, demand, attention, hiring language, and the shrinking sense that what they offer still sits near the center of the economic picture. But again, noticing is not the same as escaping. Retraining takes time. Repositioning takes time. Building new reputation takes time. Income still has to arrive while the old track weakens. This is exactly the sort of setting where the future becomes readable before it becomes easy to steer.

The same pattern appears in burnout, which is often less a dramatic collapse than a long period of visible decline. A person knows they are thinning out. They know their attention is worse, their patience shorter, their motivation flatter, their recovery slower. They know the work is costing more than it used to cost. Yet they continue, partly because stopping has consequences, partly because the outer structure still rewards endurance, and partly because nothing has yet happened that looks final enough to justify a full change. Then one day the person hits a threshold and tells the story as though the collapse came out of nowhere. It usually did not. The future was visible in weak form for a long time. The problem was not pure ignorance. The problem was that the local ability to change course was trapped inside financial needs, identity, habit, and delayed feedback.

This is what institutions do so often to human beings. They create long zones in which trouble is visible but not yet formally actionable at the level of personal comfort. They reward continuity right up until continuity becomes dangerous. They encourage people to treat endurance as wisdom, then leave them to reinterpret the damage afterward as personal failure. But much of the suffering here is structural. It comes from living inside systems where awareness and reach are badly mismatched. The person can see the weather shifting long before they can move their shelter.

That is why work is one of the clearest real life examples of this whole framework. It is not just about losing a job. It is about the long middle stage in which the future of a role begins to announce itself through weaker relevance, slower momentum, thinner trust, or growing replaceability. It is about the experience of sensing that the ground under a career has become more mobile than the language around it admits. It is about knowing that a machine is changing while still depending on the machine for rent, rhythm, and identity. It is about being asked to remain calm and productive inside a structure that may already be revising your future.

Seen this way, a lot of professional regret becomes easier to understand. People say they should have left sooner, learned sooner, prepared sooner, diversified sooner, paid attention sooner. Often they are right. But sooner would have meant acting in the stage where the signals were still deniable. It would have meant spending money, energy, or social capital before the threat had become obvious enough to feel fair. That is the hard part. Work teaches, again and again, that the future of a life can start moving before the people inside that life are granted the emotional permission to move with it. And that is exactly why careers remain one of the most familiar arenas in which visibility grows while leverage quietly dies.

Money is one of the cruelest places where this pattern shows up, because money punishes lateness with unusual precision. In love, people can still tell themselves a story. In health, people can still hope a symptom will pass. In work, people can still believe the role will stabilize. But in money, delay eventually becomes arithmetic. The numbers keep moving whether your emotions are ready or not. That is why financial life gives such a sharp view of Future Access Without Presence. People can often see pressure building long before they can stop the pressure from becoming expensive.

Most people know this feeling in ordinary form. The paycheck still arrives, but it no longer lands with the same force. Rent takes more of it. Food takes more of it. Utilities take more of it. Small surprises hurt more than they used to. The margin disappears before disaster becomes official. A person starts noticing that their life has become less forgiving. There is still income, still routine, still apparent stability, but the cushion has thinned. The future begins to feel heavier before the balance sheet looks dramatic enough to count as a crisis.

That early stage is important because it reveals something many people would rather not face. Financial trouble often becomes visible before it becomes easy to fix. The person can feel the squeeze months before they are ready to name it honestly. They know they are swiping the card with more tension. They know they are timing bills more carefully. They know they are one bad week away from a problem that would once have been manageable. They know that the same salary now buys a smaller life. But knowing this does not automatically restore leverage. In many cases it only creates a painful awareness of narrowing options.

This is one reason money shame is so powerful. People often blame themselves for not acting sooner, saving sooner, cutting sooner, exiting sooner, investing sooner, or preparing sooner. Sometimes those regrets are fair. People do make avoidable errors. People do delay for bad reasons. But a lot of financial stress is shaped by a deeper structural problem. The signals arrive before the handles do. A person can see their costs rising faster than their flexibility. They can see an unstable situation approaching while still being tied to rent, debt, dependents, contracts, habits, and the ordinary fixed costs of staying alive. They are not blind. They are constrained.

This is especially obvious in inflationary periods and cost of living pressure. People do not need an economist to tell them when daily life has become tighter. They feel it in the grocery store, in the heating bill, in transport, in insurance, in the quiet disappearance of spare room. The future shows up first as friction. A meal out becomes a calculation. Replacing something broken becomes a delay decision. A normal month begins to require low grade improvisation. The person can clearly see the direction. They know this is not temporary noise. They know the pattern is worsening their position. Yet their local power to reverse that pattern is tiny. They cannot set prices. They cannot lower national costs. They cannot force wages to catch up on demand. The problem is visible. The leverage is weak.

That is the general money version of this whole framework. Financial life often becomes more legible at exactly the moment local control becomes thinner. You can tell that housing is running away from you before you can do anything realistic to bring it back. You can see debt compounding before you can instantly erase the interest. You can sense that your savings are inadequate before income changes enough to rebuild them. You can watch a good become unaffordable before there is a clean substitute. The future is not hidden. It is right there. The problem is that visibility does not guarantee reach.

This becomes even more painful in investing and trading, because those worlds punish the confusion between prediction and control with unusual brutality. A person may be right about direction and still fail financially. That sounds absurd until you understand the timing problem. A market can move the way you expected and still move in a way that destroys you if your entry is late, your size is wrong, your risk is unmanaged, your capital is too small, or your patience runs out before the thesis matures. Being correct is not enough. In markets, the difference between seeing and steering becomes almost mathematical.

That is why so many people feel haunted by money. They often did see it. They saw the bubble. They saw the weakness. They saw that an asset was unstable. They saw that a trade had become crowded. They saw that a company was decaying. They saw that the housing move was unsustainable. But seeing the direction did not mean they had an executable edge. By the time a pattern is visible to many people, the easy part of the opportunity may already be gone. The future may be legible, but the price of reaching it has risen. Slippage, delay, risk, competition, and timing all eat the advantage. The market becomes a machine for reminding people that prediction without control is not profit.

This is part of what makes money such a uniquely educational domain. It does not flatter vague intelligence for long. You can talk brilliantly about what is coming and still be unable to convert that insight into a better outcome. You can predict a correction and lose money before it arrives. You can sense a trend and miss the part that mattered. You can know that a debt path is dangerous and still remain caught in it because the monthly structure leaves too little room for a clean turn. In money, as in so much of life, the future can be visible while the local channel of influence remains thin.

Digital finance has made this both better and worse. On one side, people have more access to information than ever. They can see prices in real time, track trends, compare products, run simulations, monitor spending, follow macro signals, and expose bad habits faster. In theory this should increase agency. Sometimes it does. Better tools can create better timing. But more often it also creates a strange overload of financial visibility without equal growth in financial control. A person can watch every fluctuation in their account while remaining stuck in the same structural pattern. They can monitor their fragility in high resolution without yet changing the conditions producing it. This is one of the darker jokes of modern money life. We have become very good at measuring our own narrowing room.

That narrowing room is where a lot of real financial suffering happens. Not the final collapse, but the long stage before it, when a person already knows the machine is becoming less forgiving. They know they are running closer to the edge. They know the debt is no longer harmless. They know the savings rate is fiction. They know the lifestyle only works if nothing goes wrong. They know a single surprise would reorder the month. That knowledge changes the nervous system. It turns money from background structure into constant atmospheric pressure. The person becomes alert, but alertness is not the same as leverage. In fact, constant financial alertness often burns energy that could have gone into the smaller, harder acts that still matter.

Those acts are usually unimpressive from the outside. Cut one recurring expense. Face the real monthly structure. Stop using credit to preserve an image of continuity. Sell the thing you keep pretending is temporary liquidity. Take the less glamorous job before pride turns into a bill. Admit that the old standard of living no longer fits the current facts. None of this feels heroic. That is part of why so many people delay it. They want a cleaner rescue, a larger fix, a more dignified solution. But money often refuses that fantasy. It forces the same lesson again and again. The useful move usually appears before it feels emotionally attractive.

There is also a collective version of this that affects entire societies. People can see financial drift long before a public turning point arrives. They can sense that wages are not matching costs, that debt is becoming normalized, that younger people are being priced out of forms of stability that once looked ordinary, that a whole class of life decisions is being delayed by basic arithmetic. These things become visible in lived experience long before they are resolved in policy or admitted in full by the culture. Once again the future leaks before the handle arrives. People know what is happening. They just do not possess a simple local lever equal to the scale of the process.

That is why money belongs so centrally in this chapter. It strips the theory down to daily reality. It shows how often a person can see the path and still be trapped inside the timing of it. It shows how often the cost of waiting is obvious before the means of escape are clean. It shows how the difference between being right and being able to act can become the difference between strain and stability, between margin and exposure, between a manageable correction and a punishing one.

In that sense, finance is not just one example among many. It is one of the clearest classrooms for this entire idea. Money teaches, with no sentimentality at all, that late clarity is expensive. It teaches that the world can become readable before it becomes reachable. It teaches that awareness can rise while flexibility falls. And it teaches that in one of the most practical areas of life, the future often arrives first as pressure, long before it arrives as an officially recognized event.

Technology has made this pattern feel almost normal, which may be one reason so many people no longer notice how strange it really is. A person opens a phone in the morning and enters a world that already seems to know them. The feed is ready. The suggestions are ready. The next clip, the next song, the next purchase, the next outrage, the next temptation, the next distraction, the next mood are all waiting in line with eerie confidence. The machine of digital life often appears to predict human attention better than humans direct it themselves. It learns habits, rhythms, weaknesses, appetites, pauses, vulnerabilities, and returns. It notices what holds the eye, what stirs the hand, what prolongs the scroll, what turns hesitation into action. It becomes very good at reading behavior, sometimes before the person doing the behavior has formed a clean thought about what they are becoming.

That is one of the most important modern examples of Future Access Without Presence. The digital world is full of structures that can see the trajectory of your attention before you can easily redirect it. A person says they only meant to check one thing, only meant to reply to one message, only meant to watch one short clip, only meant to browse for a minute, and then an hour is gone, or two, or a whole mood, or the clean shape of an evening. The future of the next hour became visible in weak form almost immediately. The person could feel the pull. They could sense the trap. They knew, often within moments, that they were no longer really choosing at full strength. But knowing that did not mean the hand regained command.

This is a very modern kind of helplessness, not dramatic, not cinematic, just incremental and intimate. It does not look like chains. It looks like convenience. It looks like frictionless design. It looks like helpful suggestion, personalized relevance, easy entertainment, smart prediction. But beneath that polished surface is a deeper fact. A great deal of digital life is built to shorten the loop by which platforms learn you, while doing very little to shorten the loop by which you regain authority over what those platforms are shaping in you. The result is an environment where prediction improves faster than personal steering.

That difference matters because attention is not a trivial thing. Attention is how time becomes life. What repeatedly captures it begins to structure mood, desire, memory, and even identity. A person may believe they are simply consuming content, but over time the content is training thresholds of boredom, tolerance for silence, appetite for novelty, emotional reflex, and the expected speed of reward. The feed is not only offering choices. It is gradually revising the conditions under which choice is felt. This is why digital drift can be so hard to interrupt. It often happens below the level of dramatic decision. A person does not wake up one day and choose a more fractured mind. They live their way into it through a thousand tiny loops in which the environment learns their direction faster than they learn how to resist it.

This is where the split between visibility and reach becomes especially sharp. Most people are not blind to what is happening. They know the phone is taking too much of them. They know the constant checking is fragmenting concentration. They know the late night scrolling is ruining sleep. They know the recommendation stream keeps giving them more of what makes them reactive, agitated, lonely, or numb. They know their own behavior has become more compulsive than free. The problem is not always lack of awareness. Very often the problem is that awareness arrives inside an environment built to absorb it without surrendering much control.

A person can know that a feed is making them worse and still open it again. They can know that short form media is thinning their patience and still reach for it in every pause. They can know that shopping prompts are training impulsive spending and still click. They can know that the constant performance of online life is exhausting them and still continue performing. This is not because they are uniquely weak. It is because digital environments are exceptionally good at turning prediction into design and design into repeated capture. They read behavior at scale, refine timing, reduce friction, and learn when a person is most open to being nudged. The person, by contrast, is often trying to defend themselves with good intentions, vague guilt, and occasional bursts of discipline. That is a very uneven contest.

There is a reason so much digital life feels like being carried slightly farther than you intended to go. One more refresh. One more clip. One more product. One more article. One more post. One more notification. Each step is small enough to seem harmless. Each step can be justified in isolation. But the trajectory appears long before the endpoint. The person often knows, early on, that they are entering a tunnel. They can feel their time slipping into a channel that will give little back. Yet the practical handle is weaker than the recognition. Exit requires effort. Friction has been engineered away for entry, not for departure. The future of the next hour becomes legible before the will to reclaim the hour remains equally strong.

This is not only about entertainment. It shapes desire itself. Platforms do not merely predict what people already want. Over time they help sculpt what feels desirable, urgent, normal, and worth noticing. A person who spends enough time in a certain digital climate begins to inherit its tempo. They become more restless, more comparative, more primed for outrage, more dependent on novelty, more unable to remain inside slow effort without craving interruption. In this way technology does not merely reflect the user. It collaborates in the production of the user. That is why the loss of steering can feel so slippery here. The environment is not only reading your present behavior. It is helping to train the future version of the person who will keep behaving that way.

This is especially obvious in recommendation culture. A person watches one thing, then another appears, then another, each one close enough to the previous desire to feel natural. The path feels self chosen because each step resembles the one before it. But resemblance is not freedom. It is often simply well calibrated momentum. The person can feel the slope. They can sense that they are being led deeper into a narrow corridor of mood, taste, politics, fantasy, fear, or consumption. Yet because each step is small, because the whole process is wrapped in pleasure or relevance, the practical moment of interruption keeps being postponed. The person says after this one, after this purchase, after this thread, after this evening, after this week. By the time they admit the pattern is real, habit has already gained strength through repetition.

There is also a social side to this that makes the whole problem heavier. Digital life turns visibility into pressure. A person does not only watch content. They watch other people being watched. They compare bodies, careers, homes, success, productivity, style, joy, and belonging through a stream of curated moments that produce silent demands. This creates a strange condition in which people can often predict their own emotional decline in real time and still struggle to step away. They know a certain app makes them feel smaller, poorer, uglier, more behind, more angry, more inadequate, more scattered. Yet the same app may also carry connection, habit, business, social proof, or the fear of missing whatever everyone else seems to be inside. Again, the future becomes visible before the exit becomes easy.

That is why digital life is one of the clearest modern examples of prediction outrunning control. The platforms can forecast attention, reaction, and appetite with extraordinary confidence because they are built from oceans of repeated behavior. The individual lives on the other side of that equation with partial self knowledge, inconsistent discipline, social dependence, and a nervous system that is not designed to remain untouched by constant low friction temptation. The environment becomes better and better at modeling the next move, while the person does not become equally better at declining it.

This has consequences far beyond screen time. It alters memory, concentration, sleep, mood, consumption, and the felt texture of daily existence. It teaches people to live in fragments. It shortens the tolerated gap between desire and reward. It makes boredom feel intolerable and silence suspicious. It can even make the slower parts of real life, reading, friendship, recovery, thought, craft, patience, begin to feel under stimulating by comparison. When that happens, the problem is no longer merely one of bad habit. The person is now living inside a feedback climate that repeatedly favors prediction over self direction.

And yet this is exactly why the example belongs here. It shows that the modern condition is not simply one of being manipulated by some distant force while remaining unaware. Much of the time people are painfully aware. They know the feed is shaping them. They know the phone is stealing from them. They know convenience has become compulsion. They know the design is not neutral. But awareness alone does not restore authority. The future of behavior can remain easy to see while the immediate path of changing that behavior remains frustratingly narrow.

Technology, then, is not just a tool in this story. It is one of the places where the story becomes impossible to ignore. It reveals a world in which environments can become astonishingly good at reading human trajectories while the humans inside those trajectories struggle to reclaim even basic forms of timing, restraint, and depth. It shows how visibility can expand while practical reach thins. It shows how a person can live inside patterns they understand and still fail to interrupt them. And it shows, with almost embarrassing clarity, that one of the signatures of modern life is not simply that we are watched, but that we are often predicted better than we are able to govern ourselves.

At the largest scale, this pattern stops feeling personal and starts feeling historical. A society can know that it is drifting before it knows how to correct its drift. A population can feel that something has become thinner, harsher, more brittle, more expensive, more exhausting, more performative, more fragmented, long before there is any clean agreement about what exactly changed or who still has the power to reverse it. That is one reason public life now feels so strange. Many people are not blind. They can see the weather of the age. What they cannot always see is a handle equal to the size of what they have recognized.

This is the scale at which Future Access Without Presence becomes almost impossible to ignore. In private life, a person may still tell themselves that what they feel is only mood, bad luck, or a local problem. But once the same atmosphere spreads across health, love, work, money, attention, and public trust, a larger picture begins to form. People start to sense that the age itself has a direction. They feel that institutions are slower than the dangers around them. They feel that public language is weaker than private recognition. They feel that society often reaches consensus only after the most useful forms of easy correction have already become costly, divisive, or late.

This happens in public health, where populations can sense that a way of living is becoming unsustainable long before collective behavior changes in proportion to the danger. It happens in housing, where entire generations can feel that access to stable life is narrowing while the mechanisms that produce that narrowing continue almost untouched. It happens in education, where people can recognize that credentials, debt, and opportunity no longer fit together in the same way, yet still find themselves moving through the old structure because the old structure remains the only recognized road. It happens in politics, where problems can be perfectly visible and still remain practically unreachable because the machinery of response is slow, fragmented, or captured by incentives that thrive on delay.

At this scale the emotional experience changes. The pain is no longer only regret. It becomes fatigue, cynicism, background anger, and that peculiar modern feeling of living inside a future that has already started while the official story has not yet caught up. People begin saying things like everyone can see this, nothing will happen until it is too late, by the time they do something the damage will already be done, this has been obvious for years. Those are not merely complaints. They are descriptions of a real structural mismatch between visibility and collective leverage.

That mismatch is intensified by the way modern information moves. Signals travel quickly now, but coordinated correction does not always travel with equal speed. A problem can become culturally visible almost overnight and still remain practically unresolved for years. A whole population can know that costs are rising, trust is falling, attention is fragmenting, family formation is changing, burnout is spreading, and public patience is thinning, yet that collective recognition does not automatically create a clean mechanism for action. Information circulates. Analysis circulates. Warnings circulate. But the social channels by which a civilization actually changes course remain slower, more contested, and more fragile than the circulation of awareness itself.

This is one reason so many people feel they are living in an era of endless diagnosis. Everyone can describe the pattern. Fewer people can still touch the gears. Public life becomes full of commentary, metrics, maps, forecasts, experts, explanations, and narratives of what is clearly happening. Yet the sheer abundance of recognition often makes the lack of response feel even more humiliating. It is one thing to suffer in ignorance. It is another to suffer while surrounded by evidence that the suffering was visible well in advance. At the social scale, this creates a kind of collective self consciousness without matching collective control.

You can see this in the language people now use about institutions. They say schools feel outdated, health care feels overloaded, work feels hollowed out, cities feel unlivable, politics feels theatrical, culture feels exhausted, the internet feels corrosive, the future feels delayed. Those phrases are often broad, but they are not empty. They reflect a common perception that systems are still moving, still operating, still producing results, but no longer doing so in a way that feels aligned with the lives inside them. The machine keeps running. The public can hear the strain in the engine. What it cannot always do is agree on where the steering wheel still is.

This is where scale becomes one of the central facts of the modern condition. Many of the systems shaping daily life are now too large, too distributed, too mediated, and too layered for any local actor to meaningfully control in the old direct sense. A person can see that housing policy, monetary policy, algorithmic design, labor market restructuring, medical burden, educational drift, and social atomization are all changing the texture of ordinary existence. But seeing that total field does not mean any single citizen, worker, parent, or voter has a simple lever proportional to it. Their awareness may expand faster than their reach, and that gap itself becomes part of the emotional structure of the age.

It is important to be exact here. This does not mean collective action is impossible. It does not mean institutions never change. It does not mean reform, organization, invention, and cultural shifts do not matter. History is full of real turning points, and societies do adapt. The point is more specific. Modern life seems to generate longer periods in which a trajectory is visible before it is tractable. It produces long preludes, long recognitions, long stretches in which people are living with an obvious pattern that remains stubbornly resistant to easy redirection. The problem is not total helplessness. The problem is delayed and diluted leverage at the scale where the problem now exists.

That helps explain a great deal of contemporary emotion. Anxiety rises when people can see where things are heading but cannot find proportionate routes for correction. Cynicism rises when official reassurance keeps arriving after trust has already eroded. Polarization rises when late action makes each possible response more expensive and more symbolic. Nostalgia rises when the past appears not only better, but simpler, as if earlier generations lived in a world where the distance between noticing a problem and acting on it was shorter. Even doom itself can be reinterpreted in this light. Sometimes doom is not pure pessimism. Sometimes it is a nervous system reacting to repeated experiences of visible trajectory without satisfying control.

This is why modern life may be defined less by ignorance than by exposure. People are exposed to signals from every level of reality. They see more of the weather of the world than earlier generations could have seen. They see macro conditions, public moods, statistical shifts, private decline, algorithmic influence, market fragility, institutional drift, and social exhaustion almost all at once. But no human nervous system was built to hold that much pattern recognition while also remaining calm about its own limited leverage. We have increased visibility faster than we have increased grounded forms of agency. That is a major reason the present feels psychologically overloaded.

In a strange way, this brings all the earlier examples together. The body shows the pattern in miniature. Relationships show it in intimacy. Work shows it in structure. Money shows it in arithmetic. Technology shows it in design. Society shows it in scale. The same underlying split keeps appearing. The future becomes readable before it remains easy to move. A trajectory enters awareness before a proportionate handle enters the hand. The result is not simply fear. It is a more complex emotion, a mixture of recognition, frustration, readiness, delay, and shame. People feel that they should be able to do more because they can already see so much. But seeing more is not the same as being placed where outcomes still answer quickly to effort.

This is the deeper reason the modern world can feel so exhausting even when many forms of life are materially safer, more connected, and more informed than in the past. The problem is not only danger. It is prolonged partial access. It is the feeling of standing near the truth without standing near the controls. It is the sensation of living in an age where warnings travel fast, but meaningful correction often travels slowly. It is the burden of knowing that many of the biggest forces shaping a life become obvious to ordinary perception long before they become responsive to ordinary action.

That does not make awareness worthless. On the contrary, awareness still matters. It matters in private choices, in earlier action, in smaller domains, in honest thresholds, in collective memory, in refusing gaslight, in organizing before the final crisis, in building local forms of contact where larger systems remain too slow. But it does change the emotional meaning of awareness. Awareness is not always a promise of power. Sometimes it is only a map of where power has already begun to thin. That is painful, but it is also clarifying. It helps explain why so many people now feel both informed and cornered, alert and delayed, awake and strangely unable to translate that wakefulness into outcomes that feel equal to what they already know.

And that may be one of the defining truths of this era. Modern life does not only produce more information. It produces more visible trajectories. It teaches people to recognize patterns early, compare them constantly, and feel the slope of events long before easy steering is available. In that sense the central wound of the age may not be simple confusion. It may be the repeated experience of clarity arriving inside systems that have already grown too large, too layered, too delayed, and too distributed to answer quickly. We do not only live in a world of uncertainty. We live in a world where the path ahead is often visible before it is reachable. That is why Future Access Without Presence no longer feels like a rare exception. It feels like one of the most ordinary realities of modern existence.

Chapter 8

There is a particular kind of pain that does not look dramatic from the outside. Nothing explodes, nothing collapses in a single visible instant, nobody blows a whistle and freezes the room. Life keeps moving in the same ordinary way it moved yesterday. The coffee is still warm, the inbox still fills, the rent is still due, the train still arrives, the body still walks around the house. And yet something inside perception has already shifted. You feel it before you can prove it. A relationship has begun to leave before anyone says goodbye. A career has begun to narrow before any official signal arrives. A body has begun to drift before the test result lands on the screen. A financial reality has begun to tighten before the numbers look bad enough to scare the people around you. You do not have certainty, but you no longer have innocence either.

This is one of the hardest human experiences to describe well, because it sounds weak when spoken too casually and melodramatic when spoken too loudly. People say they had a bad feeling, or they sensed it coming, or they knew deep down. Sometimes they are dismissed as anxious, negative, superstitious, or overly emotional. Sometimes they dismiss themselves. They tell themselves they are being ridiculous. They wait for something sharper, something undeniable, something that would make action feel justified. But the strange burden of late clarity is that the shape of the future often becomes visible before it becomes socially acceptable to name. By the time everyone agrees that the problem is real, the window for easy change may already be much smaller than it once was.

That mismatch between recognition and reach is what makes this kind of pain so corrosive. Ignorance is frustrating, but it has a bluntness to it. When you truly do not know, you cannot yet measure the distance between what you want and what remains possible. Late clarity is harsher. It lets you watch that distance grow.

It gives you enough signal to worry, enough pattern to sense direction, enough evidence to feel disturbed, but not enough standing, enough leverage, enough clean timing to simply place your hands on the world and turn it. It is not the pain of darkness. It is the pain of dim light.

Think about how often this happens in ordinary life. A couple continues to eat dinner together, continues to share bills, continues to discuss practical things, but the emotional weather has changed. The tenderness is now effortful, the jokes land slower, the silences feel heavier, the future tense stops appearing in conversation. Nothing has ended yet. No final statement has been made. No single incident is large enough to justify panic. But the arc has begun to tilt. One person may feel it long before either person is willing to name it. Months later, when the breakup finally becomes official, friends may speak as if the ending happened all at once. It rarely does. It often becomes visible before it becomes stoppable.

The same pattern appears in health, and this may be one reason health fear feels so intimate. People often notice shifts before those shifts fit neatly into formal categories. Energy changes. Sleep changes. Recovery changes. Appetite changes. Tolerance changes. A person says to themselves that something is off, but the phrase is vague enough to be ignorable. The body does not always send its warnings in clean language. It sends them as friction, as drag, as a rising cost to doing what used to feel ordinary. Later, when the clearer result arrives, people often tell the story backward, as if the decisive moment was the diagnosis. But in lived time, the diagnosis may be the moment when uncertainty ends, not the moment when the process began.

Work life produces its own version of the same wound. A person senses that their position is becoming less central, that the energy around a project has cooled, that the promises are now made in softer language, that decisions happen elsewhere, that optimism sounds rehearsed. The role still exists. The meetings still happen. The paycheck still arrives. Yet something has loosened. It is very hard to act inside that stage, because one is asked to behave as though visible drift is not real until it becomes visible enough for everyone else. The human cost is not just fear. It is the exhausting act of standing inside a future that has begun, while still being expected to speak in the grammar of the past.

What hurts in all these cases is not only the coming loss. It is the way late clarity attacks self trust. Once a person sees the pattern and still fails to alter the outcome, a cruel inner story often begins. I knew, and I did not move. I felt it, and I stayed. I saw the signs, and I still let it happen. That inner speech can become merciless very quickly, because it quietly assumes a false equation. It assumes that recognition should have guaranteed control. It assumes that seeing should have been enough to steer. It assumes that if the outcome still arrived, then the observer must have been weak, cowardly, passive, or flawed.

But that assumption is exactly what this book has been trying to break apart. There are many moments in life when knowledge and influence do not travel together. There are many moments when the future becomes more legible even as the action channel narrows. There are moments when you are not failing to act on a clear and open field. You are arriving at clarity inside a landscape whose useful windows have already been shrinking. This does not mean nothing could ever have been done. It means the timing of what could be done matters more than people like to admit.

That last point is uncomfortable, because timing feels less heroic than willpower. We like stories where the decisive factor is courage, discipline, intelligence, or conviction. We prefer narratives in which the right person steps forward, says the hard truth, makes the bold move, and changes the direction of events. Sometimes life does work that way, and those moments are real. But many other times the decisive factor is far quieter. It is whether the signal arrived early enough. It is whether the feedback loop was short enough. It is whether the person involved could still alter the trajectory before momentum hardened, before the process distributed itself across too many moving parts, before the emotional or practical cost of reversal grew too large.

This is why late clarity feels so lonely. Not only are you dealing with the possible loss itself, you are also dealing with a culture that often respects certainty more than timing. People trust declarations made after the evidence becomes undeniable. They are less comfortable with earlier forms of knowledge, because early knowledge tends to come wrapped in ambiguity. It sounds softer, less official, less defendable. You may be able to feel the curve of events without being able to produce the kind of proof that would silence every objection. So you wait. Others wait. The world keeps moving. Then one day the thing becomes obvious, and now everyone speaks as if the problem has simply appeared. In truth, the problem may have been visible in weaker form long before it became undeniable in stronger form.

There is also a bitter asymmetry here. Once an outcome finally arrives, hindsight makes the path look clean. The ending seems inevitable. The clues line up. The person who sensed it early may even be told, in a strange backhanded way, that they were right all along. But being right all along is not the same as being able to redirect the path. In fact, one of the most painful facts about human life is that correctness and power often separate at exactly the moment when a person most wants them to stay together. It is possible to be perceptive and still be late. It is possible to be honest and still be out of reach. It is possible to understand the shape of what is coming while standing too far from the controls to make understanding decisive.

Seen this way, late clarity is not merely a mood. It is a real pattern in lived experience. It is what it feels like when visibility outruns leverage. It is what it feels like when the future begins to announce itself before the practical path for changing it remains wide enough to trust. That is why this pain deserves better language than regret, weakness, or bad luck. It is not always failure of character. Sometimes it is the felt experience of arriving at knowledge inside a narrowing corridor.

And that is where any mature response must begin. Before we talk about better choices, earlier moves, cleaner feedback, or wiser forms of freedom, we have to begin by naming the wound correctly. Many people are carrying shame for outcomes that became clear only after easy steering had already begun to die. They are not merely haunted by what happened. They are haunted by the belief that seeing should have saved them. It often does not. The world can become visible before it remains reachable. That is not an excuse. It is a structure. Once you see that structure clearly, a different kind of honesty becomes possible.

Once that wound has been named correctly, the next step is not consolation, it is a better question. Most people facing uncertainty ask the wrong thing first. They ask, what is going to happen. They ask it in relationships, in money, in health, in work, in every zone of life where the future starts to gather shape before it fully arrives. The question feels natural because prediction is emotionally magnetic. It promises relief. If I can just know where this is heading, maybe I can calm down, prepare myself, defend myself, get ahead of the blow. But prediction by itself is not enough. In fact, one of the quiet traps of modern life is that people have become better and better at consuming signals about what is likely, while remaining much less disciplined about asking what can still be influenced from here.

That shift sounds small, but it changes almost everything. What is going to happen, is a forecasting question. What can still be influenced from here, is an agency question. The first one pulls your attention toward the shape of the coming wave. The second pulls your attention toward your actual contact with outcomes. The first can easily become obsession, passive scanning, or emotional rehearsal. The second forces you to locate leverage, timing, and reach. The first asks for vision. The second asks for traction.

A lot of unnecessary suffering comes from mixing those two questions together. A person senses that a friendship is cooling, that a job is decaying, that a habit is becoming dangerous, that a financial situation is tightening, and because they can feel the likely direction, they assume they are already engaged with the reality of the problem. But seeing a path is not the same as being in active contact with the points where that path can still bend. You can become a very good spectator of your own future. That is not the same thing as steering it.

This is why a more useful mental map begins with three separate zones. First, there is what is visible. These are the signals, patterns, tensions, and clues you can already observe. A change in tone, a repeated delay, a rising cost, a form of drag that keeps returning. Second, there is what is probable. This is the future shape those signals seem to indicate. Maybe the relationship is not stable, maybe the job is not safe, maybe the body is not simply tired, maybe the market is not merely volatile, maybe the family situation is not going to settle on its own. Third, and most important, there is what is still steerable. This is the zone people often neglect. These are the variables, decisions, conversations, tests, exits, commitments, boundaries, timing moves, and resource shifts that can still alter the path in a meaningful way.

Those three zones do not always overlap neatly. In fact, much of adult life consists of learning that they often come apart. Something may be visible without being easy to classify. Something may be probable without being certain. Something may be probable and visible while only a very narrow slice of it remains steerable. This is where people get trapped. They either demand certainty before acting, or they confuse awareness with influence and drift into helplessness without realizing it.

Think of a person who senses that their work life is becoming unstable. The visible zone might include meetings where their role matters less, projects that no longer have momentum, praise that sounds generic, and decisions that happen further away from them than before. The probable zone is the emerging sense that the role may shrink, stall, or disappear. But the steerable zone is different from both. It may include updating skills while there is still time, repairing a key relationship, looking for exits before the pressure becomes public, reducing financial fragility, or confronting the reality of the situation before the person is emotionally cornered. Notice what matters here. The actionable zone does not require perfect certainty. It requires honest recognition of where influence still exists.

The same pattern appears in intimate life. Suppose a couple has entered that quiet stage where the visible signals are everywhere if one is willing to see them. Less warmth, less curiosity, fewer small bids for connection, more mechanical conversation, a strange politeness where spontaneity used to live. The probable zone begins to gather around a painful thought, this may not hold. But what remains steerable is more precise. Perhaps a real conversation is still possible before resentment hardens. Perhaps counseling is still worth trying while both people still care enough to be honest. Perhaps ending things cleanly is still more humane than prolonging the drift. Perhaps a change in habit, environment, or attention could still matter, but only if it happens before indifference becomes the deeper climate. Again, the key question is not only what is happening. It is what can still be reached before the corridor narrows further.

This way of thinking is harder than it sounds because the human mind likes broad drama more than narrow leverage. We are drawn to large interpretations. Is this doomed, is this destiny, is this over, is this the beginning of collapse. Those questions feel serious. They feel like adult realism. But in many cases they are less useful than smaller ones. What decision is still live this week. What conversation has not yet become impossible. What signal have I been downgrading because it is inconvenient. What deadline is real even if nobody has announced it. What cost rises if I wait another month. What am I calling uncertainty simply because I do not want to pay the emotional price of acting early.

That last question matters because people often hide inside the word uncertainty. It sounds humble, cautious, rational. And sometimes it is. But other times uncertainty becomes a shelter for delay. A person says they are waiting for more information, when in truth they are waiting for enough permission to act without fear of regret or embarrassment. The problem is that the world rarely grants that permission on generous terms. By the time the case is obvious enough to satisfy every inner skeptic, the range of useful moves may already be thinner, more expensive, or more painful.

So the better question, what can still be influenced from here, has to be asked with discipline. It is not a slogan. It is a practice of attention. It requires looking directly at the difference between signal and wish. It requires admitting that some options exist only before they feel emotionally comfortable. It requires the humility to say, I may not know exactly how this ends, but I know enough to see that certain windows are not going to stay open forever.

There is something deeply relieving about that shift once a person truly adopts it. Not comforting, exactly, but clarifying. It cuts through the exhausting demand to predict every detail. It replaces the fantasy of total foresight with the far more practical task of identifying reachable points of intervention. You do not need to know every feature of the future to act intelligently. You need to know where your present still touches it.

This is true even in small, ordinary matters. A person notices that their sleep is getting worse, their patience thinner, their energy less stable. They may not know what larger outcome those signals point toward. But the steerable zone may already be obvious enough. Change the schedule before exhaustion becomes identity. Reduce alcohol before it becomes routine chemistry. Get blood work before fear turns vague and sticky. Stop pretending that chronic drag is a personality flaw. The same logic works in money, habits, friendships, family, health, and creative life. In each case the central question is not whether the future can be known with perfect detail. The central question is whether present action still has meaningful contact with what comes next.

A lot of maturity comes from accepting that these are different standards. Forecasting asks, how readable is the road ahead. Agency asks, where does the road still respond to my hands. Sometimes the first answer will be clearer than the second. That is one reason people can become so mentally busy while feeling practically stuck. They are spending all their energy reading, narrating, anticipating, analyzing, and rehearsing, while neglecting the narrower and more difficult work of locating leverage. Reading the future is mentally consuming. Touching it in time is often emotionally costly. Once this difference becomes real to you, you start noticing how much of life is wasted on badly framed questions. People ask whether a situation is hopeless when the better question is whether a specific intervention is still alive. They ask whether they are overreacting when the better question is whether delay improves or worsens their position. They ask whether they can prove the danger when the better question is whether the cost of waiting for proof is acceptable. They ask whether they are sure when the real issue is whether certainty would arrive too late to matter.

That is the practical turn this book has been building toward. The world is often more legible than it is reachable. If you forget that, you either blame yourself too harshly for late clarity, or you drift into prediction as a substitute for action. Neither helps. The better path begins by separating the visible, the probable, and the steerable, then asking with real honesty where influence still lives. That is not a guarantee of success. It is something more modest and more useful. It is a way of bringing your attention back to the small but real points where your life still responds.

One of the most expensive habits in human life is the habit of waiting for certainty. It sounds wise, responsible, disciplined. People are praised for being cautious, for not jumping to conclusions, for not acting too soon, for gathering more information, for making sure the case is real before they move. In some situations that caution is intelligent. It protects against panic, fantasy, and self inflicted damage. But there is another side to it, and that side is easy to miss. In many of the situations that matter most, waiting for certainty does not simply improve judgment. It quietly destroys the conditions that made timely action possible in the first place.

This is one of the hardest lessons to accept, because certainty feels morally clean. It feels like the adult standard. If I act only when the evidence is undeniable, then at least I cannot accuse myself of recklessness. If I wait until the situation is obvious, then at least I will know I was not dramatic, paranoid, or impulsive. But the world does not arrange itself around that emotional preference. It often reveals itself in gradients, not in declarations. A problem begins as a pattern before it becomes a verdict. A danger begins as drift before it becomes an event. A trajectory becomes visible in weak form before it becomes undeniable in strong form. And by the time the strong form arrives, the leverage that existed during the weak form may already be badly reduced.

You can see this everywhere once you stop pretending that clear moments always arrive with generous timing. A relationship rarely ends in the moment of the breakup itself. It often begins ending in small neglected ways long before anyone says the final sentence. The warmth thins out, the repair attempts become less sincere, the ordinary generosity of attention starts to disappear. These changes are real, but they are not yet dramatic enough to satisfy the inner judge who demands proof. So people wait. They want something decisive, something undeniable, something that will make action feel justified. They keep going, hoping for greater clarity. The problem is that greater clarity often arrives at the exact moment when the emotional structure has hardened. By then the conversation that might once have reopened the bond now becomes a postmortem. The person did get certainty. They just got it at a much worse price.

The same pattern appears in health. A body starts speaking in softer signals long before it speaks in louder ones. Sleep gets worse. Recovery slows down. Motivation sinks. Breath feels different. Energy becomes unreliable. A person notices these things, but does not yet possess a dramatic enough story to act without self doubt. So they wait for cleaner evidence. They tell themselves they are just tired, just stressed, just getting older, just overthinking it. Later, when the stronger signal arrives, they may finally move. They book the appointment, run the test, make the change. But again, certainty has not come as a gift. It has come as a bill. The earlier ambiguity was not useless noise. It was the phase in which action was psychologically hardest and often most valuable.

Careers follow the same logic. Many people can feel a decline before they can prove it. A role grows less central. A manager becomes vaguer. A project starts missing energy. The kind of inclusion that once felt natural now feels conditional. A person senses what is happening, but the evidence does not yet feel official. No one has announced anything. No formal change has been posted.

So they stay still, partly because moving early feels embarrassing. It feels like overreaction. It feels like admitting fear before fear has been socially authorized. Then one day the certainty arrives. Budget cuts, restructuring, a quiet conversation, an email, a canceled future. At that point the person often tells themselves they should have done something sooner. Usually they are right. But what they often fail to notice is that sooner would have meant acting under ambiguity, not acting under proof. This is why the demand for certainty can be so destructive. It hides the real cost of delay under the appearance of rationality. It allows a person to feel careful while the field of practical options narrows around them. It creates a strange bargain. In exchange for emotional permission, you surrender timing. In exchange for feeling justified, you become late. The mind often takes that bargain because it hates the discomfort of acting without complete validation. But reality does not care whether your move felt emotionally authorized. It only cares whether the move still had contact with the outcome.

There is a reason this problem appears so often in the human story. Acting early requires a kind of courage that does not look cinematic. It is not the courage of certainty, it is the courage of incomplete information. It is the courage to say, I do not know every detail, but I know enough about the direction to treat this as real. That is a much harder form of courage than people admit. It means risking embarrassment. It means risking that others will think you are overreacting. It means risking that your own inner voice will call you foolish, anxious, unstable, or weak. And yet many of the most important decisions in life can only be made well in that ambiguous zone. Once the case becomes obvious to everyone, the real decision may already be behind you.

This does not mean people should become frantic, impulsive, or suspicious of every weak signal. The point is not to glorify panic. The point is to understand the asymmetry between evidence and leverage. Evidence often strengthens over time. Leverage often weakens over time. Those curves do not necessarily move together. In fact, one of the central problems of adult life is that they often move in opposite directions. The proof gets better while the room to act gets worse. The story gets clearer while the road gets narrower. The pattern becomes easier to name while the cost of changing it rises.

Once you see that, a lot of familiar human regret starts to look different. People often say, I should have trusted my gut. Sometimes that is true, but the phrase can be misleading. It makes the problem sound mystical, as if the whole issue were simply one of intuition versus reason. The deeper issue is timing under uncertainty. It is not that a person possessed supernatural certainty and ignored it. It is that they had enough signal to justify attention, but not enough social or emotional permission to move without doubt. They waited for the world to become obvious, and in doing so they allowed the window for easier action to drift further away.

That is why so many painful outcomes come with a peculiar kind of hindsight. The person does not merely look back and see the signs. They look back and realize that the signs were the moment. The weak evidence was not just a prelude to the real thing. It was the actual stage at which the future was still plastic enough to be nudged. Later stages may have offered more certainty, but they often offered less reach. This is the cruel arithmetic of many turning points. Earlier means murkier. Later means clearer. But later also often means harder.

In practical life this matters enormously. The serious question is not whether you can eliminate uncertainty. You cannot. The serious question is whether you can learn to act intelligently before uncertainty disappears. That means learning to treat ambiguity not only as a nuisance, but as the normal price of timely agency. If you insist on perfect clarity before every important move, you will repeatedly pay for certainty with lost leverage. You will arrive at truth later, cleaner, and weaker.

This is true in intimate life, in health, in money, in work, in habits, in family dynamics, and in collective affairs. The world rarely sends engraved invitations to act. More often it sends patterns, weak signals, rising friction, and a sense that something is no longer moving the way it used to. Those early forms of knowledge feel unsatisfying because they do not relieve doubt. But that is exactly why they matter. They are often the zone in which freedom still exists in practical form.

A mature view of agency has to begin there. It has to accept that useful action often happens before complete certainty, not after it. It has to stop treating ambiguity as proof that nothing can be done. It has to stop pretending that waiting is neutral. Waiting is often a choice with direction. It changes the field, changes the timing, changes the cost of every later move. In many of the situations that define a life, the real danger is not ignorance. The real danger is asking for a level of proof that reality only provides after the most valuable windows have already begun to close.

This is not a tragic flaw in human thinking alone. It is a structural feature of how many processes unfold. The future often becomes more readable as it becomes less reachable. That is why certainty can be so seductive and so expensive at the same time. It offers emotional relief, but often too late. It gives you the comfort of knowing, but not always the power of changing. And once you recognize that pattern clearly, you can begin to see one of the deepest practical truths in this entire book. If you want real agency, you cannot build your life around the fantasy that the world will always become obvious before it becomes hard to move.

If timing is the hidden cost in so many failures of agency, then feedback is the hidden machinery underneath timing. People often talk about freedom as if it were a matter of inner strength, boldness, discipline, or desire. They picture agency as a property of the will, almost like a fuel that some people have more of than others. But in practice, agency depends on something far less romantic and far more concrete. It depends on whether reality is still talking back to you clearly enough, quickly enough, and honestly enough for your next move to connect with what happens next. That may sound technical at first, but it is one of the most ordinary facts in life. You can only steer what still responds to correction. You can only correct what you can still detect. You can only detect what reaches you with enough clarity before the situation has moved on. This is what feedback really means. It is not a slogan, and it is not a managerial cliché. It is the return path by which the world informs you what your contact with it is actually doing.

Without that return path, effort becomes theater. You can work harder, care more, worry more, analyze more, and still remain badly disconnected from outcomes. A person can be intensely sincere while living on delayed or distorted signals. They can be trying with all their heart, and still be trying inside a fog. That is one reason why people often feel so exhausted in situations they cannot move. They are not always lacking commitment. Often they are trapped in poor feedback conditions, where the world answers too slowly, too noisily, or too misleadingly for commitment to remain effective. Think of something as simple as driving in bad visibility. A car does not become controllable because the driver deeply wants control. It becomes controllable because the driver can see the road, feel the vehicle respond, and correct in time. If the windshield is smeared, if the steering is delayed, if the road markings vanish, if the tires grip inconsistently, then the driver does not merely become less comfortable. Their agency degrades. The same human being, with the same intelligence, the same intentions, and the same desire to stay safe, suddenly has less usable contact with the path ahead. The problem is not moral. It is informational and temporal.

The same logic governs far more of life than people admit. In relationships, poor feedback is often mistaken for mystery. One person becomes less expressive, less direct, less honest, and the other is forced to navigate through hints, silences, tone shifts, and missing warmth. They are not responding to a clean signal anymore. They are responding to fragments. This makes intelligent action much harder. It creates a strange state in which the relationship still produces emotional consequences, but no longer provides clear enough data for repair to be straightforward. People then start compensating with guesswork, anxiety, overinterpretation, or denial. The suffering grows, not only because the bond is in trouble, but because the feedback path has become unreliable.

Health follows the same pattern with brutal consistency. A body that gives clear signals can often be helped earlier. A body that gives vague, intermittent, or misleading signals is much harder to manage well. That is not because the person does not care. It is because poor feedback turns every correction into a gamble. Was that fatigue a random bad week, or the beginning of something larger. Was that pain a minor fluctuation, or a rising warning. Was that recovery lag normal stress, or a deeper drift. When the body speaks in weak or inconsistent ways, a person has to make decisions under much worse conditions. The challenge is not just lack of information. It is that the available information may arrive late, arrive blurred, or arrive mixed with noise from everything else life is already doing.

Money and work are full of the same hidden mechanics. Many people do not get clean feedback about financial risk until risk has already thickened around them. Credit can delay the felt consequence of overspending. A salary can mask the fragility of a role until the role is already near collapse. A rising market can make bad habits look intelligent for longer than they deserve. In all these cases the problem is not merely that outcomes are uncertain. The deeper problem is that feedback is delayed and distorted. A person may think things are under control because the world has not yet delivered the full cost of their current direction. By the time the signal sharpens, the room for painless correction may be much smaller.

This is why so much bad decision making is not really irrational in the cartoon sense. It is not always people acting wildly against obvious reality. More often it is people trying to manage their lives through lag. They are using numbers that are too old, impressions that are too soft, reassurance that is too comforting, or interpretations that have been filtered through hope, pride, fear, and convenience. In that condition even smart people become clumsy. Not because intelligence disappears, but because intelligence cannot do much with degraded contact.

So when we say protect feedback quality, we are talking about something much deeper than productivity advice. We are talking about defending the conditions under which agency remains real. That starts with honesty, but honesty alone is not enough. A person can sincerely want the truth and still build a life full of blurred signals. They can avoid the scale, avoid the statement, avoid the account balance, avoid the conversation, avoid the quiet evidence that would force an update. In those moments the problem is not a lack of brainpower. It is an avoidance of clean contact.

Clean contact is often emotionally unpleasant. That is one reason people avoid it. A direct health marker can be frightening. A hard conversation can destabilize a relationship that was being held together by politeness. A real financial review can puncture the story that everything will somehow sort itself out. A serious look at work can reveal that a tolerated drift has become a structural decline. But discomfort is not the same thing as damage. In fact, one of the most dangerous habits in adult life is confusing soothing feedback with good feedback. A signal is not valuable because it calms you. It is valuable because it keeps you in operational touch with reality.

This is where many forms of modern life make the problem worse. We live surrounded by streams of information, yet much of that information is too broad, too delayed, too curated, or too emotionally distorted to support real steering. People can know the mood of the internet and remain strangely ignorant of their own sleep, spending, stress, relationships, and physical drift. They can follow every major trend and still avoid the one honest metric that would tell them whether their own direction is stable. High visibility at the wrong scale can create a feeling of awareness while starving the feedback loops that matter most.

To live intelligently, then, is not just to gather more data. It is to care about the quality, timing, and relevance of the signals that shape your choices. Faster is not always better, but timely matters. More is not always better, but clearer matters. Constant monitoring is not always wiser, but direct contact matters. The aim is not to drown yourself in information. The aim is to maintain a level of contact with reality that lets correction happen before the cost multiplies.

That often means shorter loops. It means not waiting six months to look at something that is quietly deteriorating every week. It means not letting important conversations become yearly emergencies. It means checking reality in forms small enough that you do not need a full collapse to force attention. A person who only confronts the truth at crisis scale will almost always feel that life arrives too suddenly. In many cases it did not arrive suddenly at all. The signals were there, but they were too delayed, too diluted, or too avoided to support timely movement.

There is also a quieter layer to this. Good feedback is not only external. Internal models matter too. A skilled musician does not need to consciously calculate every correction because years of trained contact have compressed the loop. A good driver senses drift early. A seasoned investor, doctor, craftsperson, or parent often feels a problem sooner, not because they possess magic, but because their internal map has become better tuned to the kinds of deviations that matter. Experience, at its best, is often a way of shortening effective delay. It lets weaker signals become usable earlier. But that only helps if the person remains willing to update when reality contradicts the model. Otherwise experience hardens into arrogance, and arrogance is just another way of blinding the return path.

This is why the deepest practical form of intelligence may be less glamorous than people imagine. It is not endless confidence. It is not domination. It is not the fantasy of reading the whole future at once. It is the maintenance of contact. The person who protects feedback quality protects the conditions under which action can remain more than hope. They build lives in which reality can still answer them before the answer becomes a punishment.

Agency does not live in the volume of your intentions. It lives in the quality of your loop with the world. If that loop is clean, even modest effort can matter. If that loop is degraded, even heroic effort can arrive late. That is why protecting feedback is not a side issue in the human story. It is one of the main ways a life stays steerable while the window is still open.

There is another reason people so often feel exhausted in a world they cannot seem to move. It is not only that real control is limited. It is also that human beings are extremely good at manufacturing the feeling of control long after the substance of it has begun to disappear. This matters because fake control is not harmless. It soaks up time, emotional energy, attention, and self respect. It gives relief in the short term, but it quietly trains a person to confuse motion with influence, expression with intervention, and ritual with reach.

This confusion is everywhere. A person feels anxious about money, so they open the banking app ten times a day, recalculate the same numbers, reread the same statements, and call this responsibility. Another feels a relationship slipping, so they replay every conversation, reread old messages, monitor tone, rehearse speeches in their head, and call this working on it. Another feels the body changing, so they consume health content for hours, compare symptoms, watch videos, order supplements, and call this taking action. Another feels a career narrowing, so they obsess over industry news, complain to friends, fantasize about reinvention, and call this preparing. In each case something real is happening. Fear is real. Attention is real. Concern is real. But concern by itself is not leverage. That is one of the hardest truths in adult life. Not all movement is steering. A mind can be very busy while a life remains almost motionless at the point that matters. The body can be full of urgency while the actual channel connecting action to outcome remains untouched. This is one reason modern life feels so strangely draining. It offers endless opportunities to perform engagement without ever forcing the deeper question, did anything I did today actually change the path of the thing I claim to care about.

It helps to be brutally concrete here. Suppose a person is worried about debt. There is a version of attention that is real, and there is a version that is ceremonial. Real attention might mean making a payment plan, cutting one recurring expense, calling a creditor, moving savings automatically, or facing the full monthly structure without euphemism. Ceremonial attention might mean feeling sick every time the card is swiped, checking the balance obsessively, making grand vows at night, and then doing nothing that alters the next cycle. The emotional experience of the second version can be intense enough to feel like effort. That is why it is so deceptive. The person is suffering, and because they are suffering, they naturally assume they are engaged with the problem. But suffering is not the same as traction.

The same distinction appears in intimate life with almost cruel clarity. Many people do a great deal of emotional labor around a relationship without ever entering the narrow zone where outcomes can still be moved. They think constantly. They analyze signals. They ask friends to interpret texts. They picture confrontations, reconciliations, endings, rescues. They search for hidden certainty in small details. Yet they avoid the one direct conversation that would actually expose the structure of the bond. Why. Because direct contact is costly. It threatens identity, comfort, routine, hope. It may produce pain quickly instead of slowly. So the mind offers a substitute. It gives the person a thousand forms of symbolic control. Monitoring, interpreting, anticipating, rehearsing. The person feels active, but the relationship remains untouched at the point that matters most.

This pattern is not stupidity. It is self protection. Fake control is often emotionally attractive because it lets us discharge anxiety without paying the full price of real contact. It allows us to remain near the problem without fully entering it. It gives us something to do with our fear. That is why doomscrolling is so powerful in large scale life. It is not just information consumption. It is a ritual of simulated involvement. A person tracks every collapse, every scandal, every prediction, every warning, every trend line, and feels both informed and powerless. They experience a constant stream of activation without a matching stream of influence. The ritual becomes addictive because it keeps anxiety in motion. Motion feels better than stillness. But motion without leverage eventually turns into depletion.

There is a deep psychological comfort in symbolic action because it preserves the image of oneself as someone who cares. That image matters. Most people do not want to admit, even to themselves, that they are avoiding the actual point of intervention. So they build softer substitutes. They organize, optimize, collect, plan, read, compare, clean, prepare, announce, track, and refine. Some of those things can be useful, of course. Preparation is not always fake. Information is not always avoidance. But the dividing line is simple and ruthless. Does this behavior change the coupling between me and the outcome, or does it only change my emotional relationship to the fact that I am worried. That question can feel harsh, but it is clarifying in exactly the way this subject demands. People regularly mistake emotional discharge for practical movement. They confuse the release of tension with the movement of reality. A person writes a long angry message and does not send it. A person talks for two hours about a health scare and never books the appointment. A person spends a weekend designing a perfect productivity system and never touches the underlying work. A person reads investment material late into the night and never changes the actual risk in their finances. A person performs concern so intensely that they feel almost morally purified by it. Still, the future remains where it was.

The reason fake control is so dangerous is not only that it wastes energy. It also distorts self perception. After enough symbolic action, people begin to feel they have already tried. They tell themselves they have been dealing with the issue for months. In one sense that is true. The issue has occupied their thoughts, moods, and conversations. But in another sense the crucial work may still be untouched. This creates a nasty double burden. The person becomes both tired and unchanged. They arrive at the real point of action already depleted by the theater that came before it.

That depletion can then become another excuse for delay. Now the person says they are overwhelmed, burnt out, emotionally drained. Often they are. But part of what drained them was the long period of circular activity that never reached the lever. The problem is not merely that they failed to act. It is that they spent so much of themselves on substitutes for action that real movement now feels even more difficult than it originally was.

This is where a mature life requires a certain kind of severity, not cruelty, but severity. You have to become willing to distinguish between what soothes you and what shifts the field. You have to ask whether your rituals are reducing anxiety or increasing reach. Those are not the same achievement. A habit can be emotionally regulating and practically empty at the same time. A behavior can make you feel responsible while leaving the future untouched. A repeated check, a repeated thought, a repeated declaration, a repeated performance of seriousness can all become ways of staying busy at the edge of change instead of stepping into it.

Some of the most powerful moments in life come when a person stops negotiating with this distinction. They stop asking whether they have felt enough, thought enough, monitored enough, prepared enough, suffered enough. They ask one cleaner question. What actually changes the path from here. Sometimes the answer is small. One call. One boundary. One exit. One appointment. One reduction. One uncomfortable truth spoken clearly. One recurring expense cut. One apology made in time. One habit measured honestly. The scale is often unimpressive from the outside. That is another reason fake control wins so often. Theater feels grander than leverage. Real leverage is often embarrassingly specific.

Modern culture often pushes people toward fake control because fake control is easy to monetize, easy to broadcast, and easy to repeat. It generates content, metrics, updates, status displays, commentary, and emotional loops. Real control is quieter. It often produces less spectacle. It may even look anticlimactic. A person simply does the thing that reconnects them with consequence. They stop feeding the machine of simulation and return to contact.

That return can feel almost insulting to the ego because it reveals how little of the previous drama was actually required. Not all of it was wasted, since some forms of reflection prepare the ground for change. But much of it was overhead, emotional smoke around a lever that remained untouched. Once a person sees this clearly, they become harder to fool, including harder to fool themselves. They become less impressed by intensity for its own sake. Less impressed by noise, panic, obsessive attention, and public declarations. They begin to respect something simpler and far more valuable, the quiet act that still has real contact with the outcome.

That is why any serious response to life beyond the horizon has to include this discipline. Do not worship fake control. Do not mistake agitation for agency. Do not confuse being mentally wrapped around a problem with being structurally connected to its future. Care matters. Attention matters. Feeling matters. But only if they lead you back to the narrow point where reality still answers your hand.

The practical answer is not to dream about becoming larger than reality. It is to build a life in which your actions stay closer to consequences. Most people imagine greater agency as something dramatic, more status, more resources, more reach, more intelligence, more force. Sometimes those things help. But very often the more important change is simpler. You do not gain real freedom by standing farther from the places where outcomes are formed. You gain it by reducing the distance between what you notice, what you decide, and what happens next.

That distance matters more than people think. It is one thing to care deeply about a problem. It is another thing to live close enough to the moving parts of that problem that your correction can still matter. A person may care very much about their health, yet live in a way that turns every warning into a delayed surprise. They may care very much about money, yet arrange their habits so that the real condition of their finances is always discovered too late. They may care very much about love, work, family, and creative life, yet allow months of drift to pass before any honest contact occurs. In each case the issue is not lack of feeling. The issue is too much distance between signal and response.

This is why a more intelligent life often looks less grand than people expect. It looks like shorter loops. It looks like seeing something sooner, responding sooner, and checking sooner whether the response is doing what you hoped. It looks like removing unnecessary delay between action and correction. It looks like refusing to let important parts of life drift so far that truth only returns at crisis scale. That may not sound exciting, but it is one of the deepest practical lessons in this entire book. Real agency tends to survive where loops are shorter.

The beauty of shorter loops is that they reduce the price of honesty. If you only face reality after long periods of avoidance, every correction feels huge. The conversation becomes harder. The expense becomes heavier. The health issue becomes more frightening. The career decision becomes more disruptive. The longer the loop, the greater the emotional drama around every update. That drama then becomes one more reason to delay the next update, which makes the loop longer again. This is how people gradually build lives in which nothing is dealt with until it hurts.

A shorter loop interrupts that pattern. It makes truth smaller, more regular, and less catastrophic. A person who looks carefully at their finances every week is less likely to need a full collapse to learn where they stand. A person who speaks honestly in a relationship before silence becomes habitual is less likely to require a final scene to expose what has been dying. A person who treats bodily friction as information rather than annoyance is less likely to need a severe event before behavior changes. A person who reviews their work and direction in real time is less likely to wake up inside a life that has already narrowed around them.

This does not mean living in constant surveillance of yourself. That would simply create another form of fake control. The point is not endless monitoring. The point is timely contact. There is a difference. Endless monitoring often increases noise, while timely contact clarifies what matters. A shorter loop is not the same as obsessive checking. It is a structure that lets reality answer before the answer becomes punitive.

That idea becomes even more powerful when applied to scale. Many people suffer because they keep trying to govern outcomes that are too far away from their actual hand. They pour attention into vast situations with almost no reachable points of influence. They study massive forces, broad trends, distant dangers, collective moods, market storms, cultural shifts, and public failures, while neglecting the local conditions that still shape their own next step. The result is a painful mismatch. Their awareness grows, but their practical contact thins out. They become intellectually expanded and operationally starved.

To live closer to the horizon means becoming more honest about scale. It means asking where your hand still meets consequence. It means accepting that some domains are too distributed, too delayed, or too large for your local effort to bend in any reliable way. That acceptance is not surrender. It is discipline. It protects your remaining agency from being dissolved into grand but unreachable concerns.

This can feel humbling, even insulting, because the ego prefers large theaters. It wants to matter at impressive scale. It wants significance, range, and sweeping relevance. But life is often changed through smaller contact points. A person stabilizes their sleep. A person cuts one major expense. A person leaves one damaging pattern earlier. A person speaks one truth before resentment calcifies. A person refuses one cycle of avoidance. A person builds one routine that gives reality a cleaner way to speak back. These changes can seem almost embarrassingly modest when compared with the size of the feelings involved. Yet they are often where real leverage lives.

There is a deeper reason for this. Agency is not measured by how important a problem feels. It is measured by whether present action still couples to future outcome. That coupling is usually strongest where loops are short, signals are direct, and consequences are not endlessly postponed by noise, layers, denial, or distance. The farther you move from those conditions, the easier it becomes to confuse concern with control. The closer you move toward them, the more likely it becomes that even a small adjustment will matter.

This is why the design of an intelligent life is often the design of reachable environments. You do not need total command over the world. You need domains where your choices can still register before delay eats them alive. You need work where feedback is visible enough to support correction. You need relationships where truth can be spoken before it becomes theatrical. You need health habits that allow early detection instead of late panic. You need financial practices that expose direction before the bill arrives in full. You need routines that reduce the distance between cause and effect rather than hiding it.

All of this demands a certain kind of realism about thresholds. Every important part of life has moments beyond which things become harder to move. The body has such thresholds. Trust has such thresholds. Debt has such thresholds. Attention has such thresholds. Exhaustion has such thresholds. Meaning has such thresholds. Once crossed, they do not always make change impossible, but they often make it slower, costlier, and more painful. To live intelligently is to stop pretending that those thresholds announce themselves with perfect courtesy. They often approach quietly. That is why shorter loops matter. They let you notice movement before the threshold becomes a wall.

There is also a moral dimension to this that deserves respect. Building life closer to the horizon means refusing a fantasy that has damaged many people. The fantasy is that a good person should be able to recover from anything, at any time, with enough sincerity. That fantasy flatters the will, but it neglects timing. It ignores the fact that opportunities narrow, signals decay, and patterns harden. A mature person does not deny this. They work with it. They act before the field has fully closed. They do not call that weakness or paranoia. They call it adulthood.

This way of living also changes how you think about responsibility. Responsibility is not only about accepting blame after outcomes arrive. It is about arranging your life so that important realities can still reach you while action remains meaningful. It is about caring enough to reduce delay. It is about respecting consequence enough to stay near it. In that sense, shorter loops are not merely practical devices. They are expressions of seriousness. They say that you would rather meet truth in smaller doses than be conquered by it in one large wave.

None of this guarantees safety. A life built closer to the horizon is not a life free from loss, error, misfortune, or surprise. The world remains larger than any person. Many outcomes remain only partly steerable. But there is still a profound difference between living where your actions can register and living where they mostly arrive as commentary. That difference is one of the best names we have for practical freedom. So the task is not to become all seeing. It is not to become all powerful. It is to shorten the path between signal and response wherever you genuinely can. To reduce delay where delay is optional. To bring consequence closer where consequence has been pushed away. To stop organizing your life around late discovery. To choose smaller, cleaner forms of contact over grand but empty performances of concern. The closer you live to the places where the world still answers your hand, the more of your life remains truly your own.

In the end, the hardest thing to accept is also the most liberating. Freedom was never going to mean total command. It was never going to mean that the world would wait politely for your certainty. It was never going to mean that every danger would announce itself early, clearly, and in a form that arrived just in time for a clean correction. That fantasy is beautiful, but it has ruined a lot of people. It makes them think they failed whenever life moved faster than their proof. It makes them think they were weak whenever they saw what was coming and still could not stop it. It makes them treat timing as a personal flaw, instead of seeing that timing is part of reality itself.

A more mature idea of freedom starts somewhere humbler. It starts by giving up the dream of perfect reach. It starts by admitting that there are always going to be things larger than your grip, slower than your patience, louder than your attention, and more distributed than your local action can reliably bend. It starts by seeing that a future can become readable before it becomes touchable. It starts by understanding that knowledge and leverage are not twins. They may travel together for a while, but they are not the same thing, and they do not always stay together as events unfold.

That realization can sound cold at first. Some people hear it and think it means resignation. They think it means surrender, passivity, or a dressed up excuse for doing nothing. But that is not what it means at all. In fact, it asks more of a person, not less. It asks you to stop flattering yourself with fantasies of endless control, but it also asks you to stop insulting yourself with shame that does not belong to you. It asks you to become more exact. More honest. More disciplined about the difference between what exists, what is visible, and what is still within reach.

There is dignity in that distinction. A person who insists on total control will always be humiliated by life. A person who refuses all agency will be humiliated in a different way, by their own drift. But a person who learns to ask where contact is still alive, where timing is still favorable, where the signal is still clean enough to support a real move, that person begins to live in a way that is both smaller and stronger. Smaller in fantasy, stronger in practice. Less drunk on grand possibilities, more loyal to the narrow but real channels through which a life can still turn.

This is one reason the language of power has so often confused us. People hear power and imagine domination. They imagine force, scale, authority, size, spectacle. But much of the deepest practical freedom in life has very little to do with those things. It lives in whether your next move still meets a living consequence. It lives in whether the world can still answer you before the answer becomes punishment. It lives in whether the loop between action and reality is still short enough, clean enough, and honest enough for your effort to matter. That is not glamorous. It does not look like conquest. Often it looks like something almost embarrassingly plain. A conversation before silence hardens. A test before decline accelerates. A boundary before resentment becomes identity. A decision before exhaustion becomes structure. A truth faced early, while it is still small enough to move. This is why the answer to Future Access Without Presence is not to dream of becoming all powerful. It is to become more serious about where your actual touch still exists. It is to stop wasting yourself on the theater of concern. It is to stop confusing the intensity of your feelings with the depth of your leverage. It is to stop waiting for the world to become obvious in ways that arrive too late to help you. It is to accept that a lot of meaningful action has to happen under conditions that still contain doubt. Not total darkness, not blind panic, but ambiguity. Enough signal to move, not enough proof to feel comfortable.

That is a painful lesson, but it is also a useful one. It means that many moments people remember as failures of courage were also failures of timing. It means that some regrets can be understood more clearly, not to erase responsibility, but to make responsibility more precise. It means that the person who says, I knew something was wrong, is not always confessing irrationality or laziness. Sometimes they are describing the most normal thing in the world, the future becoming legible before the intervention channel remained strong enough to feel easy. Once you see that, compassion becomes sharper. Not softer, not sentimental, but sharper. You become less likely to moralize every late move, and more likely to ask what kind of loop the person was trapped inside, how delayed the feedback was, how narrow the reachable window had become, how much of their life had already turned into waiting for proof.

This also changes how we think about intelligence. Intelligence is not just the ability to model the future. Plenty of people can sense where things are heading. Plenty of people can describe collapse with great eloquence while standing very far from the points where collapse might still be slowed. Real intelligence includes timing. It includes respect for thresholds. It includes the discipline to act before the evidence becomes socially perfect. It includes the humility to value small reachable domains over grand unreachable theaters. It includes the courage to remain in contact with reality while reality is still willing to answer.

That answer may not always be yes. Sometimes you act and discover that the horizon had already moved farther than you hoped. Sometimes the conversation comes too late. Sometimes the body has crossed a threshold. Sometimes the market has already repriced the future. Sometimes the trust is already too damaged, the fatigue already too deep, the opportunity already too diluted. This is part of the truth too. No framework, no science, no philosophy, no discipline can erase the basic fact that life contains lateness, irreversibility, and loss. But even here the distinction still matters. There is a difference between being defeated by reality and being confused about what reality was asking of you. There is a difference between limited agency and imaginary agency. There is a difference between losing a real fight and never once touching the lever.

That difference is worth everything, because it is the difference between a life spent in fog and a life spent in contact. Contact does not guarantee victory. It guarantees something more modest and more honorable. It guarantees that your attention, your effort, and your choices are not merely circling around events after the fact. It means that you are at least trying to meet the world where it still responds. It means you are not mistaking late clarity for failure of character. It means you are learning to respect the strange arithmetic of timing, where early action often feels less justified and yet matters more, and later action feels more justified and yet matters less.

If there is any single practical wisdom hidden inside all of this, it may be this. Do not build your life around the hope that the truth will always arrive in forms that flatter your comfort. It will not. The important signals in life often come first as friction, drift, tone, lag, fatigue, hesitation, small asymmetries, tiny repeated costs. They arrive before the verdict. They arrive before the official statement. They arrive before the public version of the event. To live well is not to become paranoid about every weak pattern. It is to become skilled enough, honest enough, and brave enough to notice when weak patterns are the real beginning of strong outcomes.

That is the deeper use of this whole idea. Not despair. Not prophecy. Not mystical fatalism. Not a dramatic cult of helplessness. Just a cleaner picture of what it means to be a creature inside a world that is larger than its control, but not beyond all contact. You are not outside the game. You are not above it. You are inside it, with partial signals, delayed consequences, bounded reach, and moments of real choice that do not always look impressive when they arrive. The task is not to demand a bigger universe. The task is to stay awake to the openings that still exist in the one you have. And once you see that, the closing lesson is not bleak at all. It is severe, but it is not bleak. It tells you to act earlier, not because panic is noble, but because timing matters. It tells you to protect feedback, not because numbers and signals are sacred, but because contact matters. It tells you to abandon fake control, not because emotion is worthless, but because reality deserves better than theater. It tells you to live closer to consequence, not because smallness is holy, but because that is where a human life remains most truly alive to itself.

The future may be partly inferable without being fully touchable. That is real. The world may become obvious before it remains easy to move. That is real too. But none of that means your freedom was a lie. It means freedom was always something narrower, stranger, and more practical than people like to admit. Freedom is usable contact with reality while the window is still open.

And the deepest lesson of all is not hopelessness.

It is timing.

Epilogue

There are some ideas that arrive as inventions, and there are some that arrive as recognitions.

They do not feel new because they were absent before. They feel new because no one handed you the language for them. They were already there, living quietly inside your life, shaping your memories, bending your regrets, whispering beneath your strongest intuitions. You felt them in the silence before a breakup, in the unease before a diagnosis, in the tightening air before a market break, in the long strange season when a career, a friendship, a plan, or an entire world seemed to lean in a direction you could sense but could not stop.

This book is about one of those ideas.

It is about the brutal modern experience of seeing what is coming while lacking enough timely power to change it. It is about the gap between recognition and reach. It is about the strange wound of being neither blind nor in command. It is about the fact that the future can begin to show its outline before your hands can meaningfully alter its shape.

Most people know this feeling long before they know its name.

They know it in their bones. They know it in the body that speaks before medicine can answer cleanly. They know it in love, when the center of a bond begins to cool long before the formal ending. They know it in work, when decline becomes readable before reversal remains possible. They know it in public life, in finance, in history, in those moments when the pattern becomes undeniable and yet the machinery keeps moving as if your clarity has arrived too late to matter.

This condition deserves a name, because unnamed pain becomes self accusation.

Without language, people turn these moments into verdicts against themselves. They call it weakness. They call it cowardice.

They call it laziness, passivity, bad timing, bad character, or failure of nerve. Sometimes those accusations are true. But often something colder is happening. Often the problem is not that the person saw nothing. Often the problem is that seeing and steering are not the same thing. Often the warning is real while the leverage is already fading.

That is the territory of Future Access Without Presence.

This is not a book about prophecy. It is not a book about fatalism. It is not a book about surrender, mysticism, or magical knowledge. It is a book about structure. It is about delay, noise, friction, feedback, timing, and the harsh asymmetry between knowing and changing. It is about how the world can remain legible after it has started to become less responsive. It is about how a pattern can survive in view after your practical contact with the outcome has already begun to weaken.

That idea matters because it changes everything.

It changes how we understand regret. It changes how we understand helplessness. It changes how we understand late clarity, self blame, missed chances, and the emotional burden of modern life. It forces us to ask better questions. Not just, what is happening. Not just, who is right. But, what can still be influenced from here. Where is the window still open. Where does action still bite. Where has the signal outlived the handle.

These are not abstract questions. They live at the center of ordinary existence.

We live in an age of warnings, dashboards, feeds, indicators, forecasts, models, and endless streams of partial visibility. The world has never been so full of signals. We can often see farther than people could before us. We can detect stress earlier, drift sooner, instability faster. And yet that increased legibility has not granted equal mastery. In many parts of life, it has done the opposite. It has made the gap more painful. The map is brighter. The steering is not.

That is why this subject feels so immediate. It is not a niche idea for laboratories and specialists. It is one of the defining experiences of life in a high speed, high noise civilization. It is what it feels like to live among trajectories that reveal themselves before they become tractable. It is what it feels like to stand inside a world that keeps showing you its direction while withholding the simplicity of control.

But this is not a book of despair.

To name a limit is not to worship it. To describe a boundary is not to kneel before it. The point of this book is not to teach hopelessness. The point is to make contact with reality cleaner, earlier, and more honest. Because once you understand that prediction and power can separate, you stop waiting for certainty in places where certainty arrives after leverage decays. You stop mistaking visibility for command. You stop confusing emotional intensity with mechanical reach. You begin to act with greater respect for timing, for small windows, for signal quality, for the narrow but real channels where human agency still lives.

That is where the deeper hope of this book resides.

Not in the fantasy of total control. Not in the childish dream that enough intelligence can make a person sovereign over every system they touch. The deeper hope is more mature than that. It is the hope that clearer understanding can help you recognize what is still reachable while it is still reachable. It is the hope that better language can reduce unnecessary shame. It is the hope that once you learn the difference between seeing and steering, you begin to waste less of your life demanding miracles from moments that only ever offered timing.

So this book begins with a simple claim, and a difficult one.

There are times when the future becomes easier to read than to change.

There are times when clarity remains alive after leverage has already begun to die.

There are times when the world tells you the truth without giving you the wheel.

To understand that is not defeat.

It is the beginning of a harder, sharper, more adult kind of freedom.

END

Author's Commentary

When I started working on FAWP, I was not trying to invent a clever phrase to dress up an old intuition. I was trying to name something real. Something structured. Something measurable. Something that people feel all the time in ordinary life, but that, as far as I could tell, had never been properly isolated and described.

That something is this:

You can see what is coming, sometimes very clearly, and still be unable to do much to stop it. That gap matters. Most people know this experience in human terms long before they know it in technical language. You can feel a relationship breaking apart before the final conversation happens. You can sense a market turning before the collapse is fully visible to everybody else. You can see a health problem building, a failure building, a crisis building, and still find that your ability to intervene is weak, delayed, constrained, or already fading. The pattern becomes clearer. The forecast improves. But your usable leverage does not improve with it. Sometimes it gets worse.

That is the condition I wanted to name.

Forecasting Without Power is not about magic. It is not about prophecy. It is not about standing outside time and pulling facts out of the future like a card trick. It is about a far more ordinary and, to me, far more profound reality.

Prediction and control are not the same thing.

We often treat them as if they naturally belong together. If you knew, why did you not stop it? If the signs were there, why did nobody act? If the outcome was visible, why was it not prevented? Those questions sound reasonable, but they hide a false assumption. They assume that seeing a pattern automatically gives you force over that pattern. Real life does not work like that. Some systems become easier to read precisely when they are becoming harder to move.

That is the heart of FAWP.

What pushed me toward this idea was not just philosophy. It was the repeated sense that modern life is flooded with information while remaining strangely poor in timely influence. We have alerts, dashboards, models, forecasts, analytics, warnings, signals, and endless streams of measurement. In theory, this should make us more capable. Sometimes it does. But sometimes all it does is make helplessness more detailed.

You do not just suffer the event. You suffer the long corridor leading toward it. You watch it gather shape. You understand the direction. You may even explain it correctly in advance. And still, the machinery keeps moving.

That is why FAWP matters to me as more than a technical idea. It is a human one. But I also came to believe it is more than a metaphor. I believe FAWP points toward a real, newly identified physical and information-theoretic phenomenon, one that appears when predictive access and actionable influence decouple inside delayed, noisy, constrained systems.

That last part is important.

I am not saying the universe has suddenly broken its own rules. I am saying the opposite. FAWP appears because the rules are being obeyed. In embedded systems, access is limited. Signals arrive with delay. Noise corrupts what is known. Action does not pass instantly into outcome. Influence weakens with latency and constraint. But predictive structure can survive longer than effective steering. In other words, the future can remain partially inferable after it has stopped being meaningfully alterable from the point of view of the embedded agent.

That is not mysticism. That is structure.

At the research level, the idea became clear when I separated two things that are usually blurred together. The first is predictive coupling. That means the present still contains genuine information about what happens later. The second is steering coupling. That means an action taken now still has detectable power to shape what happens later.

Those are different channels.

A great deal of ordinary confusion comes from treating them as one. But once they are separated, an entirely new regime becomes visible. Predictive coupling can remain alive after steering coupling has effectively collapsed. Signal can survive after leverage dies. Awareness can remain after influence drops below operational relevance.

That is FAWP in its cleanest form.

From there, the work stopped feeling like a loose philosophical reflection and started feeling like the discovery of a real regime. Not just a mood. Not just a metaphor. A regime with boundaries, measurements, thresholds, and tests. A regime that can be simulated, analyzed, and falsified. A regime where the future becomes easier to read than to move.

To me, that is the most exciting part.

FAWP suggests that one of the most common frustrations in modern life is not merely psychological. It may reflect a genuine structural feature of embedded existence. There are situations where the forecast is real, the pattern is real, the warning is real, and yet the available intervention channel is too delayed, too weak, too noisy, or too constrained to matter in time. In those situations, the problem is not ignorance. The problem is operational access.

That distinction changes how we think.

It changes how we talk about responsibility. It changes how we judge failures. It changes how we understand institutions, systems, and even personal experience. Not every preventable-looking disaster was truly preventable at the moment it became visible. Not every correct warning arrived with real leverage attached to it.

Sometimes knowledge rises while force decays. Sometimes visibility increases inside a structure that is already slipping out of reach.

That is why I think FAWP deserves to be treated seriously.

The phrase may sound simple, but the implications are not. If the phenomenon is real, then many of our habits of judgment are too crude. We ask who knew, but not who could still act. We ask whether a signal existed, but not whether the action channel was still open. We ask why the forecast failed to stop the event, when the deeper question may be whether the system had already crossed into a regime where forecasting and power had come apart.

That is the conversation I want this work to open.

I am not presenting FAWP as a final answer to everything. I am presenting it as a new lens, and I believe it is a powerful one. It gives language to an experience that is widespread but poorly named. It gives structure to a gap that many people feel but cannot easily explain. And it suggests that beneath this familiar human frustration there may be a deeper scientific truth: under the right conditions, prediction can outlive influence.

That, to me, is the real significance of FAWP.

It says that the future is not simply hidden until it arrives. Sometimes it leaks into the present as pattern before it remains open as action. Sometimes signal survives where leverage does not. Sometimes the world lets you read more than it lets you move.

And if that is true, then FAWP is not just a useful phrase.

It is the beginning of a discovery.

Coda: The Age of Warnings

We live in an age of warnings.

That may be one of the simplest ways to describe modern life.

Everywhere you look, something is flashing, updating, predicting, or advising. There are alerts on phones, charts on screens, risk scores in reports, trend lines in news stories, health markers in apps, forecasts in markets, models in research, and endless little signals telling us that something is shifting, building, drifting, or about to give way. The world feels less silent than it used to. It feels annotated. Tagged. Measured. Interpreted in real time.

And yet, for all this visibility, people do not feel more powerful.

That is the strange part.

You would think more warning would naturally produce more control. If we can see more, surely we can act better. If we can detect trouble early, surely we can prevent it. If the signs are visible, surely the outcome is still open. That is the promise hidden inside the modern worship of information. More data will save us. More monitoring will protect us. More awareness will lead to better timing, better choices, better futures. Sometimes it does. But not always.

Sometimes warnings arrive inside conditions where useful leverage is already thin. Sometimes the signal gets clearer while the action window gets smaller. Sometimes the world becomes easier to read at exactly the moment it becomes harder to move.

That is why FAWP matters.

Forecasting Without Power is not just a technical idea. It is a description of one of the defining tensions of our era. We are surrounded by signs, yet often starved of force. We know earlier, track better, and measure more precisely than people in the past ever could, but that does not mean we can redirect events at the moment that matters. In many situations, the warning channel is strong while the intervention channel is weak, delayed, or clogged with friction.

That is not a contradiction. It is a condition.

You can see this everywhere once you learn how to look for it.

A person can tell their own body is heading toward exhaustion long before collapse. The signs are all there. Sleep changes. Mood changes. Focus breaks down. The warning is real. But the pressures shaping daily life do not vanish just because the pattern becomes visible. Awareness rises, but room to act may remain narrow.

A family can feel trouble building long before anyone speaks the final sentence out loud. The air changes. The pauses get longer. The warmth thins out. Future outcomes become easier to predict. Yet that increase in clarity does not guarantee repair. Sometimes recognition comes after too much momentum has already built.

A market can look increasingly unstable while the people inside it remain unable to guide where it goes next. A city can watch strain accumulate in housing, water, transport, or health. A company can see a problem forming in plain sight. In each case, warning is not the same thing as command. Seeing danger does not mean holding the steering wheel.

This is one of the reasons modern life can feel so psychologically exhausting. It is not only that bad things happen. It is that they often arrive with a long prelude of visibility. We do not simply endure the event. We endure the approach. We watch shape emerge from noise. We recognize the pattern before the impact. We know enough to worry, but not enough to change the outcome with confidence. That prolonged interval between recognition and influence is part of what makes the age of warnings feel so heavy.

It is a world full of signals without guarantees.

And that has consequences for how we think about intelligence, responsibility, and blame.

When something goes wrong, people often ask: Did nobody see this coming?

But that is not always the right question.

Very often, somebody did see it coming. Sometimes many people did. The deeper question is whether the warning arrived while meaningful action was still possible. Was there time? Was there reach? Was there enough access, enough coordination, enough usable leverage to alter the path? Or did the warning become clear only after the larger arrangement had already gathered too much inertia? That distinction matters.

Without it, we confuse awareness with control. We assume that knowledge should automatically translate into force. But the whole point of FAWP is that these are different things. Prediction and influence do not travel together by default. They can separate. In fact, under delay, noise, friction, and constraint, they often do.

That separation may be one of the most important hidden facts of embedded life.

We are not outside the world looking down on it from some clean, godlike distance. We are inside it. We receive fragments. We act through lag. We work through bottlenecks, partial information, conflicting incentives, and limited reach. Our forecasts are made from within the flow, not above it. That means a future can become partially legible before it remains operationally touchable. The signs may arrive, but not the handle.

This is why an age of warnings does not automatically become an age of prevention.

A warning is only as useful as the path between seeing and doing.

If that path is slow, corrupted, expensive, politically blocked, physically delayed, or buried under noise, then even accurate recognition may fail to produce meaningful change. The world may tell you what is coming without giving you enough force to stop it.

It may reveal direction without surrendering control. It may let you read ahead while keeping your hands tied to the present.

There is something almost cruel about that, but it is also clarifying.

It tells us that more measurement alone is not the same as more agency. More dashboards do not guarantee more power. More alerts do not guarantee better timing. More prediction does not guarantee control. In some cases, it only makes the gap more visible. It turns helplessness into a chart. It gives powerlessness a timestamp.

That is not a reason to reject warning. It is a reason to understand it properly.

Warnings matter. Forecasts matter. Signals matter. They are often the only reason we know the shape of danger at all. But if we want to live honestly in an age of warnings, we have to stop pretending that visibility by itself is enough. We have to ask harder questions. Not just what was known, but when it became actionable. Not just whether signs existed, but whether leverage still existed when the signs became clear. Not just whether the future could be read, but whether it could still be moved.

That is the deeper lesson of FAWP.

The future is not always hidden until it arrives. Sometimes it announces itself early. Sometimes it leaks into the present as pattern, pressure, and probability. Sometimes the world whispers before it strikes. But the whisper is not the same thing as permission. A warning is not a guarantee of control. Clarity is not command.

We live surrounded by forecasts, but that does not mean we live surrounded by power.

And perhaps that is the real signature of our time.

Not silence, but signal.

Not ignorance, but asymmetry.

Not blindness, but the growing realization that the world may let us see more than it lets us change. That is the age of warnings.

Coda: Where Power Ends

Human beings love a comforting myth.

We love the idea that if we can only see clearly enough, then we can act clearly enough. If the truth is visible, then change should still be possible. If the signs are present, then the outcome should still be open. In this picture, knowledge and power travel together. Awareness naturally becomes leverage. Vision naturally becomes control. It is an attractive story.

It is also incomplete.

One of the hardest things to accept about life inside real systems is that clarity and force do not always arrive at the same time. Sometimes they do. Sometimes a warning comes early enough to matter, a decision reaches the right place, and a small correction changes the future before it hardens. Those are the moments we celebrate, and rightly so. They make us feel that intelligence is useful, that foresight has dignity, that paying attention is rewarded.

But not every situation works that way.

Some situations have inertia. Some accumulate momentum. Some move through delay, friction, noise, and scale until they reach a point where the structure becomes easier to read even as it becomes harder to move. The direction sharpens. The pattern emerges. The likely outcome stops feeling vague and starts feeling obvious. Yet the handle you would need to change it is already slipping away. That is where power ends.

Not in total darkness. Not in perfect ignorance. Not in the absence of signal. Quite often, power ends in the presence of growing visibility. The tragic thing is not that nothing can be seen. The tragic thing is that seeing may arrive after the last easy opening for influence has already begun to close. That possibility unsettles us because it challenges one of our deepest moral habits. We are used to thinking that if a person understood enough, they could still do enough. We are trained to ask: if you saw it coming, why did you not stop it? Why did you not leave sooner, act faster, warn louder, push harder? The question sounds practical, but hidden inside it is a fantasy that knowledge always comes with reach. FAWP asks us to let go of that fantasy.

It asks us to consider that prediction and influence are not identical powers. They are different channels. One concerns what can be inferred from the structure of the present. The other concerns what can still be changed through action. In a simple, responsive world, they may line up closely. But in delayed, noisy, constrained systems, they can separate. Prediction can remain alive after actionable leverage has already begun to collapse.

That is not mystical. It is not an excuse. It is not surrender. It is a description of a real condition. There are moments when the future is no longer hidden, but not yet stoppable.

That condition appears across scales. A person can feel their life drifting toward exhaustion long before they find a clean way to step out of the machinery causing it. A family can sense a fracture widening long before any member has the force, timing, or shared ground needed to repair it. A society can detect instability building long before it can coordinate a meaningful correction. In each case, the forecast may become clearer while the intervention path grows thinner.

This is why the end of power rarely feels dramatic in the moment. It does not always announce itself as a slammed door. More often it feels like drag. Delay. Resistance. A strange lag between recognition and effect. You speak, but nothing moves. You decide, but the outcome does not bend. You understand more and more, yet the world answers with less and less room.

That narrowing is one of the hidden experiences FAWP is trying to name.

It is the experience of reaching the boundary where awareness continues but leverage does not. And that boundary matters because it changes how we understand both ourselves and the systems around us. If power can end before visibility does, then a great deal of human frustration begins to make more sense. The problem is not always that people are blind. Sometimes they are seeing perfectly well. The problem is that the world has already moved into a regime where signal survives longer than control. That is a hard truth, but it is also a clarifying one.

It tells us that not every failure should be interpreted as stupidity. Not every missed prevention should be treated as proof that nobody understood. Not every correct warning was delivered inside a window where effective force still existed. Sometimes the forecast was accurate. Sometimes the pattern was real. Sometimes the future was visible. What was missing was not intelligence, but the operational ability to turn recognition into consequence.

That distinction does not erase responsibility. But it deepens it. It forces us to ask better questions. Not only who knew, but who could still act. Not only whether danger was visible, but whether leverage was still available when visibility became sharp. Not only whether a signal existed, but whether the channel between decision and effect had already degraded beyond useful range.

Those are sobering questions, because they move us away from fantasy and toward structure. And once we begin thinking structurally, the world looks different.

We begin to notice that modern life is full of systems that are easier to monitor than to redirect. We become very good at measuring, tracking, forecasting, and modeling. We fill the world with indicators. We place confidence intervals around uncertainty. We convert risk into dashboards and warnings into notifications. Yet the ability to act in time does not automatically expand alongside this flood of information. In some cases, the opposite happens. The system becomes more legible while the individual inside it becomes less effective.

That is the deeper reason FAWP matters.

It is not merely a theory about prediction. It is a theory about the uneven relationship between knowledge and force in embedded life. It reminds us that awareness does not guarantee reach. It reminds us that seeing the road ahead is not the same as holding the wheel. It reminds us that there are boundaries, sometimes quiet and sometimes brutal, beyond which the future may still be readable but no longer easily movable.

That is where power ends.

But this is not a purely pessimistic conclusion. There is a kind of wisdom in recognizing the boundary. The point of naming it is not to glorify helplessness. The point is to become more honest about timing, access, and leverage. If we understand that prediction and control can decouple, then we stop mistaking visibility for victory. We stop assuming that measurement is the same thing as agency. We begin to ask earlier, harder, more practical questions about when intervention still matters and what conditions allow it to remain real.

In that sense, the boundary is not only an ending. It is also a lesson.

It teaches humility about what knowledge can do by itself. It teaches respect for delay, friction, and scale. It teaches us that power is not a permanent companion of intelligence. It is a local and perishable condition. It depends on timing. It depends on access. It depends on whether the path from action to outcome is still open when the signal arrives.

And perhaps that is the final point.

The future does not always wait for us to catch up to it. Sometimes it leaves traces before it arrives. Sometimes it becomes partially visible while still resisting our hands. Sometimes the world lets us understand more than it lets us change.

That is not prophecy. That is not fate. That is the shape of life inside constrained systems.

And if FAWP has done anything worthwhile, I hope it has given a name to that shape.

Not so we can admire it from a distance, but so we can recognize, with greater honesty, where power truly lives, how it fades, and what it means when understanding continues after leverage begins to run out.

Coda: The Future Is Not Silent

One of the oldest habits of the human mind is to imagine the future as something sealed.

A closed door. A dark room. A blank page.

We speak as if tomorrow does not exist until it arrives, as if events come into being only at the moment they strike us. The future, in this picture, is mute until it becomes present. It gives no signal. It offers no outline. It keeps its face hidden and then steps forward all at once.

But that is not how life usually feels.

In lived experience, the future is often not silent at all.

It rustles. It leans. It gathers weight. It leaves traces in advance.

A relationship rarely vanishes in one clean instant. Long before the final sentence is spoken, something in the air begins to change. The timing changes. The pauses change. The effort changes. A person may not know every detail of what is coming, but they feel direction before arrival. The future starts to make noise before it becomes fact.

The same is true in larger systems. A market under strain often does not fail out of nowhere. Pressure builds. Volatility shifts. Confidence wobbles. Small signs begin to align into a larger pattern. A body heading toward collapse often speaks before it falls. Sleep drifts. focus weakens. Mood flattens. A city under pressure starts murmuring before it breaks. Delays lengthen. faults repeat. stress concentrates in familiar places.

Again and again, the same truth appears.

The future is not silent.

It does not always announce itself clearly. It does not hand over a schedule. It does not explain its own meaning in plain language. Sometimes it whispers in fragments. Sometimes it shows itself only through pressure, rhythm, drift, or repetition. But it often leaves enough structure in the present that something attentive can begin to sense where things are heading.

That is one of the deepest intuitions behind FAWP.

Forecasting Without Power begins from a simple but powerful observation: the future can become partially readable before it becomes meaningfully alterable. The signs can arrive before the handle. The pattern can emerge before the power to redirect it does. In some cases, the cruelest thing is not that events strike without warning. It is that warning exists, but usable leverage does not keep pace with it.

That is why the idea matters.

If the future were truly silent, then ignorance would be the whole story. We would simply be surprised, again and again, by a world that refuses all advance structure. But that is not the world most people live in. Most people know the opposite feeling. They know the long lead-up. They know the uneasy interval where something has not happened yet, but has already started to feel real. They know what it is to see a shape forming in the distance and feel, with growing certainty, that its direction is not good.

The problem is that seeing is not the same as stopping.

That is the fracture FAWP tries to name.

The future may speak before it arrives, but it does not always speak inside a window where action still matters. That is the painful asymmetry. Signal can strengthen while leverage weakens. Clarity can increase while control drains away. The path from recognition to intervention can be slowed by delay, cluttered by noise, narrowed by friction, or blocked by scale. In those moments, the world is not hiding what comes next. It is simply not surrendering enough force to change it.

This is why so many experiences of modern life feel haunted by a strange mixture of knowledge and helplessness.

We live among indicators. We are surrounded by warnings, forecasts, dashboards, probabilities, metrics, and continuous updates. We know earlier than ever. We model more than ever. We track change in real time. And yet, for all this visibility, people often feel less effective than the abundance of information seems to promise.

That is not a contradiction once FAWP comes into view.

If the future is not silent, then more signals should not surprise us. But if power does not automatically travel with those signals, then more knowledge will not always produce more control. Sometimes it only produces a sharper experience of the gap. You can hear the future approaching without being able to step far enough aside.

There is something almost merciless about that condition.

Not because it is mystical, but because it is structural.

The future speaks through the present because systems carry momentum. Causes accumulate. constraints tighten. trajectories harden. The present is not cut off from what comes next. It already contains hints, pressures, and directional residues. That is why prediction is possible at all. But prediction does not mean mastery. To extract signal from a process is not the same thing as having enough reach to bend the process in time.

That distinction changes how we think about failure.

When bad outcomes happen, people often ask whether the signs were there. Usually, that question is treated like a moral test. If the signs existed, then somebody should have acted. If the warning was visible, then prevention should have followed. But real systems are harder than that. Sometimes the signs are absolutely there. Sometimes the forecast is not even the problem. The real issue is that the channel between seeing and doing has already become too weak, too delayed, too narrow, or too expensive to carry enough force.

In other words, the warning may have been real, but the reach was not.

This is one reason FAWP belongs not only to theory, but to human experience. It names the feeling of being inside a world that tells you more than it lets you change. It gives language to that long, frustrating zone where tomorrow has already begun to speak through today, yet your hands still fail to connect with the larger motion.

And once you recognize that, many things look different.

You become less tempted by the fantasy that awareness always equals agency. You become less eager to confuse visibility with command. You begin to ask better questions. Not just what was known, but when it became actionable. Not just whether a signal existed, but whether the route from signal to effect was still open. Not just whether the future left traces, but whether those traces arrived in time to matter.

Those are harder questions. They are also more honest ones.

Because the future does not owe us silence.

It often speaks early. It leaks into the present through pattern, strain, drift, and probability. It reveals itself not as certainty, but as shape. Sometimes that shape is faint. Sometimes it is unmistakable. But in either case, the important lesson is the same: reality is not waiting passively to become legible only at the moment of impact. It is often legible in advance, at least in part.

That is not prophecy. It is not magic. It is not a claim that human beings can stand outside time and inspect what has not happened yet. It is simply the recognition that the present already carries structure, and that structure can point forward before it becomes fully actual in experience.

That is why the future is not silent.

And perhaps that is both the comfort and the cruelty.

The comfort is that the world is not pure chaos. It leaves traces. It has continuity. It carries direction. There is signal in the flow of things. The cruelty is that hearing those signals does not guarantee that we can answer them with enough force. Sometimes the future speaks before it arrives, and all we gain from that speech is awareness without leverage.

Still, even that matters.

To name the condition is already to see more clearly. To understand that signal and control can separate is already to think more honestly about what embedded life is like. We are not blind, but neither are we sovereign. We live inside processes that can become readable before they remain movable. We receive warnings from within the stream, not from above it.

So the final lesson is not that the future is hidden.

It is that the future often speaks in advance, and that one of the great challenges of being human is learning how to live inside that speech, how to distinguish signal from noise, and how to recognize, with as much honesty as we can, the difference between hearing what is coming and still having the power to change it.

About the Author

Ralph Clayton is an independent writer and researcher whose work explores one of the oldest questions in human life through a modern lens: what does it mean to know something is coming, yet still be unable to change it?

His writing lives at the intersection of physics, information theory, complex systems, and lived human experience. Rather than treating science as something distant or sealed off from ordinary life, Clayton focuses on the places where abstract ideas become deeply personal: delay, uncertainty, limited access, failed timing, and the strange gap between seeing and doing. His work asks how prediction, agency, and constraint interact inside real systems, whether those systems are physical, social, psychological, or economic.

At the center of this research is **Forecasting Without Power**, or **FAWP**, a concept Clayton developed to describe the condition in which the future becomes easier to predict than to influence. The idea gives language to a familiar but often overlooked experience: recognizing the shape of what is coming while lacking the leverage to meaningfully redirect it. Through this framework, Clayton examines how signal, action, and timing can separate, and why the modern world often gives people more warning without necessarily giving them more control.

Clayton's broader theoretical work also includes the **Volumetric Time Model**, a framework that distinguishes between what exists, what is accessible, and what can be operationally influenced from within an embedded point of view. Across his writing, he is drawn to a recurring theme: reality may be fully structured, yet our access to that structure is always partial, delayed, noisy, and constrained. In that tension, he sees not only a scientific problem, but a human one.

What makes Clayton's work distinctive is his effort to bridge formal reasoning and everyday recognition. He is interested not only in equations, models, and simulations, but in the ordinary experiences that make abstract ideas feel real: sensing a market turn before a crash, feeling a relationship fail before the final conversation, or recognizing a developing crisis before the machinery around it can respond. His goal is to build language for those moments without mystifying them, and to show that many of the most frustrating experiences of modern life may reflect deeper structural truths.

As both a theorist and a writer, Clayton is committed to making difficult ideas accessible without flattening them. His style combines scientific seriousness with clarity, intuitive metaphor, and a strong interest in the emotional reality of uncertainty. He writes for readers who want big ideas that still connect to the world they actually live in.

In this work, Ralph Clayton invites readers to think more carefully about prediction, power, timing, and the boundaries of action. His central question is simple, but profound:

What happens when the future speaks before we can still move it?

Don't miss out!

Visit the website below and you can sign up to receive emails whenever Ralph Clayton publishes a new book. There's no charge and no obligation.

https://books2read.com/r/B-A-WZAFF-TNTCJ

BOOKS 2 READ

Connecting independent readers to independent writers.

Did you love *Forecasting Without Power: Why the Future Can Be Visible Before It Is Changeable - Deluxe Edition*? Then you should read *The Volumetric Time Model: Why the Future Feels Decided - Deluxe Edition*[1] by Ralph Clayton!

[2]

THE VOLUMETRIC TIME MODEL*Why the Future Feels Decided***Deluxe Edition**

What if your life isn't something that unfolds moment by moment, but something that already exists as a complete shape?

Not a script. Not fate. A structure.

In **The Volumetric Time Model**, Ralph Clayton invites readers to step outside the familiar idea of time as a ticking clock and see it the way modern physics quietly does: as a volume. A

1. https://books2read.com/u/b6NoDM

2. https://books2read.com/u/b6NoDM

three-dimensional object containing every possible path your life could take, from birth to death, all existing at once.

From that perspective, the past doesn't disappear. The future isn't empty. And the strange feeling that "this was always going to happen" starts to make sense.

Written for general readers with no physics background, this book blends modern ideas from relativity, information theory, and complexity science with everyday experiences we all recognize. The feeling of predicting an outcome but being unable to change it. Watching events slide toward inevitability despite your best efforts. Knowing what's coming, yet lacking the leverage to stop it.

Clayton introduces a clear, intuitive framework for understanding these moments. In the Volumetric Time Model, reality is globally consistent, but access to it is local and delayed. You move through a pre-existing landscape of possibilities without ever seeing the whole terrain. Prediction can remain accurate even after influence collapses. You can see the curve of the road long after the steering wheel stops responding.

This **Deluxe Edition** expands the original work with **new material and visual depth**, including:

• **Three additional codas (Coda I, II, and III)** that extend the book's central ideas beyond the final chapter• **An expanded closing section** that reframes the model in human, philosophical, and practical terms• **Additional original illustrations** designed to make the geometry of time, access, and agency visible rather than abstract• A refined reading flow that rewards both first-time readers and those returning to the model

Across eleven chapters, **The Volumetric Time Model** rethinks free will, causality, responsibility, and choice, not as metaphysical riddles, but as engineering problems shaped by access, latency, noise, and information flow. Agency appears not as a binary state, but as a fragile channel that can narrow, distort, or vanish entirely without anyone noticing.

Along the way, the book explores:

• Why the future can feel "decided" without being predetermined• How prediction can persist even when control is impossible• Why modern systems remove leverage quietly rather than issuing commands• How responsibility depends on access, not hindsight• What it would actually take to disprove this model of time

The tone is conversational, visual, and human. Complex ideas are explained through metaphors, stories, and lived experience rather than heavy mathematics. For readers who want to go deeper, each chapter concludes with a concise Formula Deck that outlines the underlying structure without interrupting the narrative.

This is not a book about catastrophe. Nothing explodes. Nothing breaks. The lights stay on. Life continues. That continuity is the clue.

The Volumetric Time Model: Deluxe Edition is for readers who sense that something fundamental has shifted in how decisions land and how the future arrives. It offers a calm, unsettling, and ultimately clarifying explanation for why seeing what's coming no longer guarantees the power to change it.

You are still moving forward.

Now you can finally see the shape you're moving through.

Read more at https://www.ralphclayton.uk/.

Also by Ralph Clayton

The Eterra Cycle
The Last Architect of Eterra: Book One of the Eterra Cycle - Deluxe Edition

The Modern Covenant Deluxed
The Religion of Almost Made It: A Covenant for Lost Souls - Deluxe Edition

The Spectacle Trilogy - A Precursor Cycle
The Imaginary Pot of Vodka Gold - Final Cut
Tales from the Motherland — Book Zero

The Temporary Arrangements Series
How to Be Nothing by the Majestic Man
The Children of Kings: A Study in Useful Men - Deluxe Edition
Please Remain Seated: This Will Only Take a Moment:
How Hunger Is Measured: A Childhood - Deluxe Edition
You May Now Proceed: Nothing Is Stopping You - Deluxe Edition

You May Now Proceed Nothing Is Stopping You

Standalone

Tales From The Motherland - Deluxe Cut

The Volumetric Time Model: Why the Future Feels Decided - Deluxe Edition

Forecasting Without Power: Why the Future Can Be Visible Before It Is Changeable - Deluxe Edition

Watch for more at https://www.ralphclayton.uk/.

www.ingramcontent.com/pod-product-compliance
Lightning Source LLC
LaVergne TN
LVHW090312210926
848480LV00001B/219